Advanced Technologies in Earth Sciences

Editors:

U. Münch, Germany
L. Stroink, Germany
V. Mosbrugger, Germany
G. Wefer, Germany

For further volumes:
http://www.springer.com/series/8384

Michael Kühn • Ute Münch
Editors

CLEAN

CO_2 Large-Scale Enhanced Gas Recovery in the Altmark Natural Gas Field – GEOTECHNOLOGIEN Science Report No. 19

Springer

Editors
Michael Kühn
GFZ - German Research Centre
for Geosciences
Centre for CO_2 Storage
Potsdam
Germany

Ute Münch
GEOTECHNOLOGIEN Coordination Office
Potsdam
Germany

ISSN 2190-1635 ISSN 2190-1643 (electronic)
ISBN 978-3-642-43345-0 ISBN 978-3-642-31677-7 (eBook)
DOI 10.1007/978-3-642-31677-7
Springer Heidelberg New York Dordrecht London

Printed on acid-free paper

Springer is part of Springer Science+Business Media (www.springer.com)

Foreword

Advanced Technologies in Earth Sciences

Ute Münch, Ludwig Stroink, Volker Mosbrugger, Gerold Wefer (Eds.)
"Advanced Technologies in Earth Sciences" is based in the German Geoscientific Research and Development Programme "GEOTECHNOLOGIEN" funded by the Federal Ministry of Education and Research (BMBF) and the German Research Foundation (DFG).

This programme comprises a nationwide network of transdisciplinary research projects and incorporates numerous universities, non-university research institutions and companies.

The books in this series deal with research results from different innovative geoscientific research areas, interlinking a broad spectrum of disciplines with a view to documenting System Earth as a whole, including its various sub-systems and cycles. The research topics are predefined to meet scientific, socio-political and economic demands for the future.

Preface

Germany has ambitious goals to protect the climate from anthropogenic carbon dioxide (CO_2) emissions as agreed in Kyoto in 1997. This is because CO_2 is held responsible by the majority of scientists for a significant part of the actual global warming. In Germany, an enormous amount of the energy provision is still achieved through, e.g., coal-fired power plants, known for their huge emissions of carbon dioxide. While renewable energy sources are becoming well established, the question remains if the transition to renewables can take place fast enough to largely reduce our greenhouse gas emissions? However, it is to be expected that fossil fuels will still be needed for decades to come. In particular coal is cheap and abundant and will be therefore a significant part of the energy portfolio worldwide in the midterm future. This holds especially true for countries like Brazil, Russia, India, China, South Africa (the so-called BRICS countries), and many more. One solution to the problem is decarbonisation of our coal-based power grid in combination with CCS-technology (Carbon Capture and Storage).

Since 2005, the Federal Ministry of Education and Research (BMBF) of Germany has promoted scientific projects for the development of such technologies in the framework of the Research and Development programme GEOTECHNOLOGIEN. Two GEOTECHNOLOGIEN Science Reports Vol. 6 and Vol. 14 have been published on this topic so far.

Many scientific, technological and safety aspects were investigated since then with regard to CO_2 storage. In recent years, much has been learnt on petrophysical and mineralogical properties of cap rocks like salt or mudstones and from the storage formations, as a base for studies on CO_2-induced alteration e.g. mineral dissolution and precipitation processes. Furthermore CO_2 propagation and fluid replacement processes in the storage complexes and with its gas fluxes in different rocks and formation layers have been modelled with existing and newly developed simulation codes for visualization purposes and to deepen the process understanding. Geochemical and geophysical monitoring techniques have been tested concerning their feasibility to control CO_2 injection and migration and to ensure a safe and long-term storage of CO_2. Various drilling, wellbore assessment and abandonment technologies have been investigated. In addition, microbiological activities in the deep biosphere and their reaction to CO_2 were studied.

The CLEAN project was planned as a second pilot site in Germany on a research scale to gain experience in this case with enhanced gas recovery (EGR) through the active injection of CO_2 with less than 100,000 t in total into a depleted gas reservoir. Although the permission for the CO_2 storage was not granted by the mining authorities within the time frame of the CLEAN project, many valuable experiments were conducted and excellent results obtained which are highlighted in this Science Report. In that way, the technological, logistic and conceptual prerequisites are provided for implementing a CO_2-based EGR project in the Altmark, and a benchmark is set for similar projects in the world.

Geological storage of CO_2 offers the opportunity to keep enormous amounts of CO_2 permanently out of the atmosphere and the carbon cycle, which are otherwise released from coal-based power generation, from energy-consuming industrial processes (steel, cement) and from burning biomass. To be able to use this innovative technology on an industrial scale in the midterm, more efforts need to be undertaken by governments, scientists and industrial partners. Particular attention should be paid to the permanent and safe storage of CO_2. This focus and projects like CLEAN are prerequisites for the effective mitigation of the climate and the public perception of such new technologies.

Ute Münch
Head of the GEOTECHNOLOGIEN coordination office
Michael Kühn
Project coordinator CLEAN

Contents

Chapter 1
Introduction to the Joint Research Project CLEAN: *C*O_2 *L*arge-Scale *E*nhanced Gas Recovery in the *A*ltmark *N*atural Gas Field

Michael Kühn, Frank R. Schilling and Heinz Wendel

Abstract The joint project CLEAN was a Research & Development (R&D) action with its scientific programme accompanying a pilot Enhanced Gas Recovery (EGR) project designed by GDF SUEZ E&P Deutschland GmbH (GDF SUEZ) in cooperation with Vattenfall Europe. It was funded by Germanys Federal Ministry of Education and Research (BMBF) in the period from 1 July 2008 to 31 December 2011. Within the framework of this R&D project a total of 16 German scientific and economic institutions participated.

The project was set up as pilot to investigate the processes relevant to EGR by the injection of CO_2 into a subfield of the almost depleted Altmark natural gas field (Germany). Despite the setback that permission for active injection was not granted by the mining authority during the period of the project, important results fostering the understanding of processes linked with EGR were achieved.

The CLEAN results provide the technological, logistic and conceptual prerequisites for implementing a CO_2-based EGR project in the Altmark and provide a benchmark for similar projects in the world.

M. Kühn (✉)
Helmholtz Centre Potsdam, GFZ – German Research Centre for Geosciences, Centre for CO_2 Storage, Telegrafenberg, 14473 Potsdam, Germany
e-mail: michael.kuehn@gfz-potsdam.de

F.R. Schilling
Helmholtz Centre Potsdam, GFZ – German Research Centre for Geosciences, Umweltgeotechnik, Telegrafenberg, 14473 Potsdam, Germany

now at KIT Karlsruher Institute of Technology, Institute for Applied Geosciences, Adenauerring 20b, 76131 Karlsruhe, Germany

H. Wendel
GDF SUEZ E&P Deutschland GmbH, Waldstrasse 39, 49808 Lingen, Germany

M. Kühn and U. Münch (eds.), *CLEAN: CO_2 Large-Scale Enhanced Gas Recovery in the Altmark Natural Gas Field*, Advanced Technologies in Earth Sciences,
DOI 10.1007/978-3-642-31677-7_1,

CLEAN was a scientific programme in support of a pilot Enhanced Gas Recovery (EGR) project which was designed to be carried out in cooperation with GDF SUEZ E&P Deutschland GmbH (GDF SUEZ) and Vattenfall Europe (Kühn et al. 2009, 2011, 2012). Within the framework of this R&D project a total of 16 German scientific and economic institutions participated (Fig. 1.1).

The Altensalzwedel block, a separate geological section of the Altmark natural gas field, which is hydraulically isolated due to its structure and is almost completely exploited, had been selected by the field operator for this purpose (Fig. 1.2). The operator also developed and provided the infrastructure required for the implementation of the pilot project above and below ground. The aim of the project CLEAN was to develop and test technologies, which enable to store for a long time significant amounts of CO_2 economically, ecologically and safe in a depleted natural gas reservoir. Hence the main objectives of the joint research project were:

1. To test whether it is possible to mobilize gas quantities which cannot be recovered by conventional means in order to enhance gas production and
2. To study the natural gas field in view to its general suitability for the injection and storage of CO_2 and whether the CO_2 injection is generally possible.

Research in the framwork of CLEAN mainly focused on the development and optimization of suitable techniques and methods: (1) for an optimum CO_2 injection with the purpose to exploit the gas reservoir to its maximum extent (*Chap. 2 – Technical Aspects of CO_2 Injection*); (2) to determine and assess the integrity of existing wells, including test of a long-term well closure (*Chap. 3 – Well Integrity*); (3) to study, describe and evaluate all processes relevant to CO_2 injection and the displacement of natural gas in the reservoir (*Chap. 4 – Evaluation of Geo-processes*); (4) to monitor the entire geological complex from the surface down to the reservoir, with regard to CO_2 propagation as well as focus on environmentally relevant processes (*Chap. 5 – Reservoir and Cap Rock Monitoring and Chap. 6 – Environmental and*

Fig. 1.1 Partners from academia and industry in the project CLEAN. In addition to GFZ- German Research Centre for Geosciences and GDF SUEZ (Gaz de France) the consortium includes Bundesanstalt für Geowissenschaften und Rohstoffe (BGR) in Hannover, Christian-Albrechts-Universität Kiel (CAU), DBI-Gastechnologisches Institut GmbH in Freiberg, DMT (Deutsche Montan Technologie) in Essen, Eberhard-Karls-Universität Tübingen, Freie Universität Berlin, Friedrich- Schiller-Universität Jena, GICON Großmann Ingenieur Consult GmbH in Dresden, Helmholtz-Zentrum für Umweltforschung GmbH (UFZ) in Leipzig, IFINKOR – Institut für Instandhaltung und Korrosionsschutztechnik gGmbH in Iserlohn, Martin-Luther-Universität Halle-Wittenberg (MLU), Technische Universität Clausthal (TUC), Technische Universität Dresden and Friedrich- Alexander-Universität Erlangen-Nürnberg

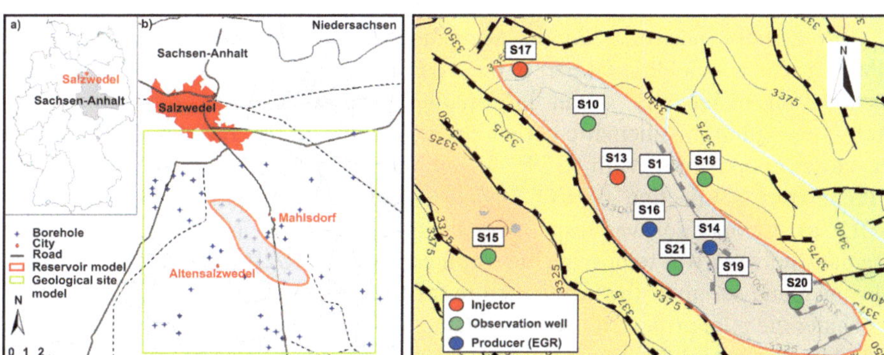

Fig. 1.2 Location of the Almark natural gas field in Germany with the Altensalzwedel block south east of the town Salzwedel (*left*). The reservoir model outlined in *red* indicates the study area. The enlargement of the Altensalzwedel block shows the various injector, producer and observation wells (*right*)

Fig. 1.3 Organization chart of the joint R&D project CLEAN (compare Fig. 1.1 for abbreviations)

Process Monitoring); and (5) to provide extensive information to the public and other stakeholder (*Chap. 8 – Public Outreach*).

The joint research project was coordinated by the Helmholtz Centre Potsdam, GFZ – German Research Centre for Geosciences in close cooperation with all research partners and the field operator GDF SUEZ. The core of the joint research project consisted of three thematic networks: "Well Integrity", "Evaluation of Geo-Process" as well as "Environmental and Process Monitoring" (Fig. 1.3). Each thematic network comprised four projects. They in turn were the framework for further sub-projects. Irrespective of the given structure of CLEAN, the sub-projects were also cross-linked to each other throughout the entire project.

The thematic network "Public Acceptance" which had also been funded by BMBF became an additional part of the R&D project. The objective was to provide factual and technically sound information especially to the regional public. Activities within this thematic network were conducted by GFZ and implemented in close cooperation with GDF SUEZ and BMBF. The thematic network "Technikum – Injection of CO_2" was associated with CLEAN as well. GDF SUEZ as site operator was responsible for this part of the project. Purpose of this thematic network was to provide the technical basis for R&D within CLEAN which also included set-up of the infrastructure for CO_2 injection.

Under the pilot project, it had been planned to inject nearly 100,000 t of CO_2 into the Altensalzwedel block. However, the permitting procedure required under mining law was delayed by significant political influence as well as the delays in transposing the Directive of the European Parliament and Council on geological storage of carbon dioxide (DIRECTIVE 2009/31/EC) into German federal law. Therefore the responsible mining authority (State Office for Geology and Mining of

Saxony Anhalt) did not issue a permit for CO_2 injection during the period of the project.

Despite this setback for the CLEAN project, it was possible to complete most of the scientific work. This was made possible by adjustments to the implementation and the objectives. The thematic networks "Well Integrity" and "Evaluation of Geo-Process" were hardly affected and were thus able to reach their targets. The basic projects within the thematic network "Environmental and Process Monitoring" were realised. Exception was the project "Reservoir- & Cap Rock-Monitoring" because it was closely linked to the not permitted CO_2 injection. However, it was possible to successfully test all methods developed at other sites (Groß Schönebeck, Windischeschenbach).

In the light of the general political conditions, which were still not clear in the second quarter of 2010, GDF SUEZ had to postpone any further orders for services to subcontractors. The opening of the information centre at Salzwedel which was ready for work and fully equipped is still lacking. Public relations were adapted to this situation and continued. As an alternative, the information material that had been developed is presented to the public at the GFZ and the Ketzin visitor centre (http://www.co2ketzin.de).

Work carried out under the joint research project CLEAN led to a comprehensive evaluation of the EGR potential of the Altmark field. Digital databases and models were developed (compare Chap. 4) and the well integrity, potential production as well as consequences and risks of the EGR technology at the Altmark site were compiled (compare Chap. 3). It was also possible to document, summarize and assess the major in-situ processes occurring in connection with CO_2 injection in the depleted gas reservoir (compare Chap. 4). The development and testing of different monitoring techniques facilitates improved surveys of CO_2 storage sites (compare Chaps. 5 and 6). This not only increased the knowledge about EGR and CO_2 storage but as well of the deep underground in general (e.g. geothermal reservoirs, exploration of hydrocarbons). In addition to that, the results of the joint research project provide the technological, logistic and conceptual preconditions for implementing an EGR pilot project with CO_2 injection at the Altmark site. This conclusion is underlined as well by the "Theoretical test case of the injection of 100 kt of CO_2 into the Altmark depleted gas field" (compare Chap. 7). Feasibility of such pilot injection is further emphasized by results from Jenkins et al. (2012) who state that safe storage and effective monitoring of CO_2 in depleted gas fields is possible based on their results from the Otway site in Australia.

The optimized use of the potential subsurface storage capacity and the reduction of risks during future EGR and CO_2 storage projects are the main economic benefits of the work carried out. However, the economic success of the accomplished work will largely depend on the political developments over the next few years and decades.

Acknowledgements We owe many thanks to the authors and participants of the CLEAN project who contributed to this volume. We would also like to express our gratitude to the project managers Dr. Maja Tesmer (GFZ) and Dr. Peter Pilz (GFZ) who made CLEAN a success.

In the name of all participants and partners of the project CLEAN we wish to thank the German Federal Ministry for Education and Research (BMBF) for the financial support (Grants 03G0704A, B, C, E, G, H, J, K, M, N, O, P, Q, R, S, U, V, W, X) within the Research & Development Programme GEOTECHNOLOGIEN and the Project-Executing Agency (PTJ) at Rostock for very good and professional support.

Our sincere thanks go to all project partners for the efficient cooperation and to GDF SUEZ as site owner for initiating this project and their general support.

References

Jenkins CR, Cook PJ, Ennis-King J, Undershultz J, Boreham C, Dance T, de Caritat P, Etheridge DM, Freifeld BM, Hortle A, Kirste D, Paterson L, Pevzner R, Schacht U, Sharma S, Stalker L, Urosevic M (2012) Safe storage and effective monitoring of CO_2 in depleted gas fields. Proc Nat Acad Sci USA 109(2):E35–E41. doi:10.1073/pnas.1107255108

Kühn M, Tesmer M, Meyer R, CLEAN-Partner (2009) CLEAN – Ein F&E-Verbundvorhaben zur Erhöhung des Ausbringegrades von gasförmigen Kohlenwasserstoffen durch die Injektion von Kohlenstoffdioxid (CO_2). In: Stroink L, Gerling JP, Kühn M, Schilling FR (Eds.) Die dauerhafte geologische Speicherung von CO_2 in Deutschland – Aktuelle Forschungsergebnisse und Perspektiven. GEOTECHNOLOGIEN Science Report no. 14. Koordinierungsbüro GEOTECHNOLOGIEN, Potsdam, pp 112–118 (http://www.geotechnologien.de)

Kühn M, Förster A, Großmann J, Meyer R, Reinicke K, Schäfer D, Wendel H (2011) CLEAN: Preparing for a CO_2-based enhanced gas recovery in a depleted gas field in Germany. Energy Procedia 4:5520–5526. doi:10.1016/j.egypro.2011.02.538

Kühn M, Tesmer M, Pilz P, Meyer R, Reinicke K, Förster A, Kolditz O, Schäfer D, CLEAN Partners (2012) CLEAN: *CO_2 Large-scale Enhanced gas recovery in the Altmark Natural gas field (Germany)*: project overview. Environ Earth Sci. 67:311–321 doi:10.1007/s12665-012-1714-z

Chapter 2
Technical Aspects of CO_2 Injection

Robert Meyer, Gernot Voigtländer and Heinz Wendel

Abstract The almost depleted Altmark gas field was chosen by the owner and operator GDF SUEZ E&P Deutschland GmbH (GDF SUEZ) for an Enhanced Gas Recovery (EGR) project. GDF SUEZ took part in the joint research project CLEAN providing this site as a basis for the scientific work of the partners from academia and industry. In November 2007, GDF SUEZ filed the application for injection of up to 100,000 t of CO_2 with the State Office for Geology and Mining of Saxony Anhalt. The permitting process came to a halt towards the end of 2008, because the responsible mining authority considered a national CCS (Carbon Capture and Storage) law to be the only legal basis for approval. The national CCS law still has not been enacted in Germany. In January 2009, the erection of the interim CO_2 storage and conditioning unit in Maxdorf was completed. New flow lines between the Maxdorf and the potential injection wells were fully planned but never build. Corrosion resistant re-completion of the injection wells did not take place either. Within the funding period of the CLEAN project (2008–2011), there was no injection of CO_2.

The Altmark natural gas field located in the federal state of Saxony-Anhalt (Fig. 2.1), owned and operated by GDF SUEZ E&P Deutschland GmbH (GDF SUEZ) is known as the second largest gas field in onshore Europe (area of ca. 1,000 km^2). Production started in 1969 and reached top peak production in the mid 1980th with approximately 12 billion cubic metres per year (Fig. 2.2). Today, after production of approximately 265 billion cubic metres gas, the field is in its tail end phase (average recovery of total gas in place is 80 %) with an annual production of some 500 million cubic metres. Due to the gas production, the initial reservoir pressure is reduced from initially 42.5 MPa to approximately 8.5 MPa.

R. Meyer (✉) • G. Voigtländer • H. Wendel
GDF SUEZ E&P Deutschland GmbH Waldstrasse 39, 49808 Lingen, Germany
e-mail: Robert.Meyer@gdfsuezep.de

M. Kühn and U. Münch (eds.), *CLEAN: CO_2 Large-Scale Enhanced Gas Recovery in the Altmark Natural Gas Field*, Advanced Technologies in Earth Sciences,
DOI 10.1007/978-3-642-31677-7_2,

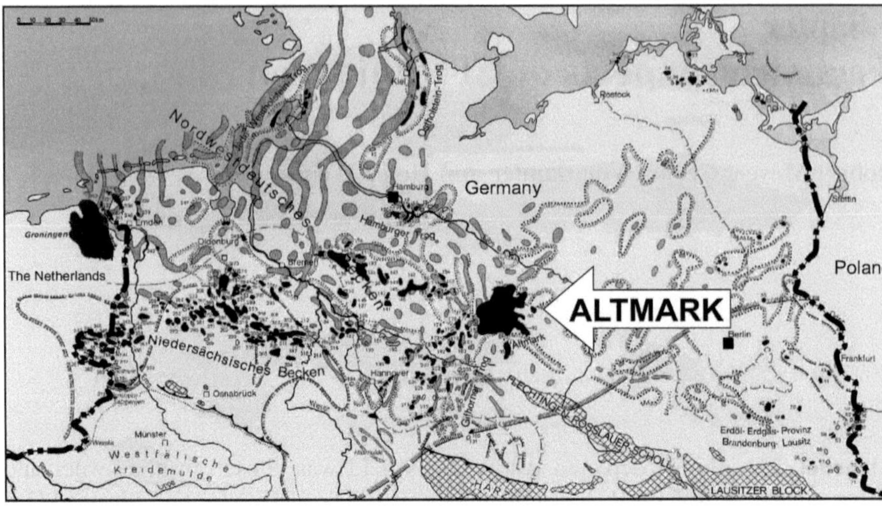

Fig. 2.1 Location of the Altmark natural gas field. Gas fields (*dark grey*), oil fields (*grey*) and underground saline structures (*light grey*) (Source: LBEG Lower Saxony)

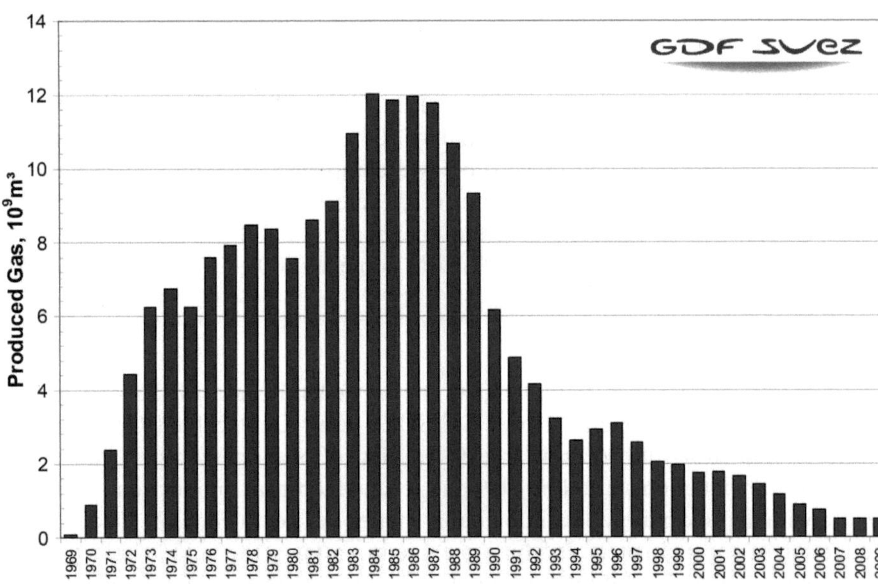

Fig. 2.2 Annual production of natural gas from the Altmark field between 1969 and 2009 (in billion cubic metres)

In Germany, research activities dedicated to CO_2 storage as potential means to mitigate climate change started more than 10 years ago at the Federal Institute for Geosciences and Natural Resources (BGR) and the GFZ – German Research Centre

Fig. 2.3 The interim-storage and conditioning unit near Maxdorf (Altmark). In the very front of the two 300 m^3 vessels, two injection pumps are installed. Four air cooled evaporators can be seen to the *left hand side*

for Geosciences. Besides saline aquifers in the North German Basin (Gerling et al. 2009), the Altmark, as the only almost depleted giant onshore gas field with an assumed storage potential of approximately 500 million tons of CO_2 (May et al. 2003), attracted attention from researchers as well as the power generating industry.

In this context and with regard to further utilisation of the Altmark gas reservoir, GDF SUEZ was approached by several industry partners since 2003. In 2006, GDF SUEZ started conceptual arrangements towards a future pilot project to be conducted in the structurally and hydraulically isolated subfield Altensalzwedel, which is representative for the entire Altmark field, with regard to the reservoir properties and infrastructure in place. In September 2007, GDF SUEZ and Vattenfall Europe (VE) agreed to commission the pilot project for Enhanced Gas Recovery (EGR) in cooperation.

During preparation of the project, the Federal Ministry of Education and Research (BMBF) invited GDF SUEZ to take part in the national joint research project CLEAN. Therefore, in November 2007, GDF SUEZ commenced the application procedure for the EGR pilot project to receive permission for CO_2 injection from the mining authority in Saxony-Anhalt, the State Office for Geology and Mining, under the given mining law. The application covered the installation and commissioning of surface facilities (Fig. 2.3) for taking-over and conditioning the

CO_2 provided by VE via trucks from the Schwarze Pumpe pilot power plant southeast of Berlin. In addition, the relaying of appropriate flow lines and recompletion of wells chosen for CO_2 injection were key activities proposed. In March 2008, the CLEAN project was approved by the BMBF and started in the following July. In January 2009, the CO_2 storage and conditioning unit in Maxdorf was erected. However, by the end of 2008, the approval process came to an unexpected halt. Despite the fact that the application was filed under mining law in the first instance, the mining authority considered a national CCS (Carbon Capture and Storage) law to be the only legal basis to approve the injection permission from that time on (per October 2011, the national CCS law still has not been enacted).

Injection of up to 100,000 t of CO_2 in the framework of the EGR pilot project was supposed to provide the technical basis for the joint R&D project CLEAN. Aim was to test, evaluate and approve the general capability of the Altmark reservoirs for EGR and CO_2 storage as well as to demonstrate the technical capability to condition and inject CO_2 into Altmark Rotliegend reservoir. Further, it was planned to elaborate and evaluate future opportunities for an economically attractive utilization of the gas field.

The following steps were planned to be tested:

- Take-over of liquid CO_2 transported by trucks from the pilot plant Schwarze Pumpe to the Altmark;
- Interim storage of CO_2 in two 300 m^3 vessels at pressure and temperature of 1.5 MPa and 27 °C, respectively;
- Conditioning of CO_2 in order to test different aggregate states and to run different injection regimes with regard to volumes and rates, including the optimisation of the entire process with regard to energy consumption;
- Routing the CO_2 to the two injection wells at a maximum pressure of 9.0 MPa in a temperature range from +8 °C up to +40 °C;
- Injection of CO_2 into the Rotliegend reservoir (depth >3,000 m, temperature ca. 120 °C) and thereby testing the well performance.

Technical specifications of the CO_2 infrastructure comprising the conditioning unit at Maxdorf, the flow lines to the bore holes and the injection wells in the field were planned to be or even built as described in the following sections.

The CO_2 storage and conditioning unit at Maxdorf consisted of:

- Two unloading stations for the trucks;
- Two vessels of 300 m^3 volume each, to store the liquid CO_2;
- Two pairs of air cooled evaporators to be used during CO_2 injection for keeping constant pressure on the vessels by vaporization of liquid CO_2;
- Two booster pumps;
- Two high pressure pumps;
- A pre-heating system allowing for different flow regimes;
- A condensing unit to catch the boil off gas from the vessels and to pump the liquid CO_2 back into the vessels;
- A manifold for four flow line connections.

For transportation of the CO_2 from the conditioning unit in Maxdorf to the potential injection wells, flow lines were planned but not built:

- Two new flow lines (6 in. nominal diameter) were thought to be linked and connected to the potential injection wells with pipelines of 770 and 3,380 m length, respectively;
- The chosen material for the flow lines was stainless steel (ferritic-austenitic) AF-22 in order to be able to maintain the conditions required for liquid-, gaseous- or supercritical CO_2;
- Pipelines were planned to be equipped with cathodic corrosion protection.

For the purpose of CO_2 injection the potential two wells would have required a re-completion to build in corrosion resistant materials. For well heads and tubings it was planned to use Cr^{13}-steel.

Within the funding period of the CLEAN project between July 2008 and December 2011, there was no injection of CO_2. However, despite this setback for the CLEAN project, it was possible to complete most of the work although some adjustments to the implementation and the objectives were necessary.

References

Gerling JP, Reinhold K, Knopf S (2009) Speicherpotenziale für CO_2 in Deutschland. In: Stroink L, Gerling JP, Kühn M, Schilling FR (Eds.) Die dauerhafte geologische Speicherung von CO_2 in Deutschland – Aktuelle Forschungsergebnisse und Perspektiven. GEOTECHNOLOGIEN Science Report no. 14. Koordinierungsbüro GEOTECHNOLOGIEN, Potsdam, pp 28–38 (http://www.geotechnologien.de)

May F, Brune S, Gerling JP, Krull P (2003) Möglichkeiten zur untertägigen Speicherung von CO_2 in Deutschland – eine Bestandsaufnahme. Geotechnik 26(3):162–172

Chapter 3
Well Integrity

Kurt M. Reinicke, Mingxing Bai, Peter Bock, Carsten Fichter, Zhengmeng Hou, Ralph Krebs, Hans-Jürgen Kretzschmar, Udo Lubenau, Robert Meyer, Mohsen Rafiee, Rico Rockmann, Meinhard Schmidt, Günter Schmitt, Steffen Schmitz, Catalin Teodoriu, Martina J. Weichmann, Falk H. Weinlich, Patrick Were, Lars Wundram and Yucheng Zhang

Abstract The implementation of an underground CO_2 storage requires evidence that the storage is and will remain tight in the future. This refers to the cap rock and the wells penetrating it. Assessment and verification of well integrity of accessible wells is technically possible. The available methods, allowing a direct assessment, were evaluated and a measurement and testing strategy is proposed. Unlike accessible wells, already plugged ones require predictive methods for their assessment. These are based on well information and a comprehensive understanding of the coupled thermal, hydraulic, mechanical and chemical processes during well construction, operation and after abandonment. The methods have been applied to a well zone characterised by conditions typical for the subsurface in the area of interest and with regard to the potential injection site in the Altmark. The calculated safety margins emphasize that technical well integrity of the 12 examined boreholes

K.M. Reinicke • M. Bai • C. Fichter • Z. Hou • C. Teodoriu • M.J. Weichmann (✉) • F.H. Weinlich • P. Were • L. Wundram
Institute of Petroleum Engineering (ITE), Clausthal University of Technology, 38678 Clausthal-Zellerfeld, Germany
e-mail: martina.weichmann@tu-clausthal.de

P. Bock • H.-J. Kretzschmar • U. Lubenau • M. Rafiee • R. Rockmann • S. Schmitz
DBI Gastrechnologisches Institut gGmbH, 09599 Freiberg, Germany

R. Krebs
Institute of Petroleum Engineering (ITE), Clausthal University of Technology, 38678 Clausthal-Zellerfeld, Germany

Currently at DBE TECHNOLOGY GmbH, 31224 Peine, Germany

R. Meyer • M. Schmidt
GDF SUEZ E&P Deutschland GmbH, Waldstrasse 39, 49808 Lingen, Germany

G. Schmitt • Y. Zhang
IFINKOR-Institut für Instandhaltung und Korrosionsschutztechnik gGmbH, Kalkofen 4, 58638 Iserlohn, Germany

M. Kühn and U. Münch (eds.), *CLEAN: CO_2 Large-Scale Enhanced Gas Recovery in the Altmark Natural Gas Field*, Advanced Technologies in Earth Sciences,
DOI 10.1007/978-3-642-31677-7_3, © Springer-Verlag Berlin Heidelberg 2013

is given for enhanced gas recovery injecting 100,000 t of CO_2 in the Altensalzwedel subfield without a need for any further intervention.

Self-healing of defects was investigated in full-scale experiments under in-situ conditions. In addition to the expected self-healing as a result of salt creep, healing was also observed resulting from the interaction of salt, cement and casing with dry or wet CO_2.

Methods and technologies for CO_2 well monitoring and intervention presented here are sufficient for the mining safety of CO_2 storage wells under high pressure. The system contains technologies proven under field conditions as well as procedures in which CO_2 was applied.

An innovative well abandonment concept was developed and tested in the field for the long-term containment of CO_2 in depleted Rotliegend gas reservoirs. It aims at amending the conventional standard well abandonment procedure, takes advantage of the natural creeping ability of the thick, homogeneous Zechstein salt formation located at around 3,000 m depth in the Altmark area and consists of four main sealing units: (1) a standard sealing element with cement from the reservoir to the impermeable cap rock, (2) a salt plug created in the formerly reamed section of casing within the plastic Zechstein (Upper Permian) rock salt formation, (3) two bridge plugs at the bottom and top of the salt plug and (4) standard cement sealing elements from the top bridge plug to the ground surface. Comprehensive numerical simulations conducted prior to and during the field test in 2010 and 2011 successfully predicted the evolution of the now proven convergence using downhole measurement data. This new long-term sealing concept has been successfully tested at the Altmark natural gas field.

3.1 Well Condition

A safe and ecological underground storage of CO_2 requires first of all the long-term technical integrity of the wells affected by the injected CO_2. In line with other investigators, technical integrity is assumed if there is no significant leak in the subsurface system, i.e. the casing, the tubing, or the packer, and if there is no significant fluid movement from the CO_2 storage into an underground source of drinking water. Evaluation of actual well integrity is relatively simple for accessible wells, since they may be surveyed to directly assess the conditions of the wells. The evaluation of the integrity of abandoned wells or the evaluation of well integrity development over longer time frames can only be done indirectly and require relevant information on geology and wells, and a comprehensive understanding of relevant THMC (thermal–hydraulic–mechanical–chemical) processes affecting well integrity. Relevant information for the North German Basin was collected, evaluated, and documented in several publications (Tran Viet 2010; Reinicke and Weichmann 2011). The information was further specified for the area under consideration in the CLEAN project. The potential storage reservoir is a depleted gas field at a depth of more than 3,000 m with pressures significantly lower than hydrostatic. The reservoir is capped by massive Zechstein salt layers of several

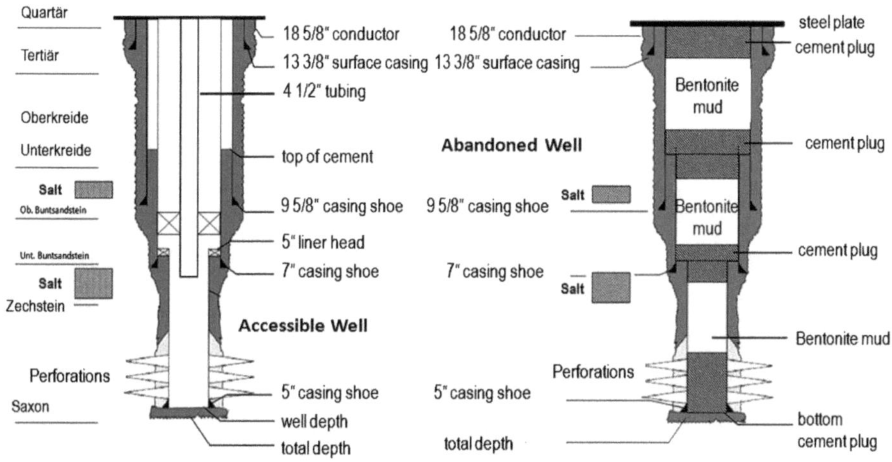

Fig. 3.1 (**a**) Current well standard. (**b**) Current abandonment standard (Reinicke and Fichter 2010a)

hundred meters thickness. Wells, penetrating the potential storage reservoir, were identified to have been drilled in the late 1960s and later. Out of the 420 wells penetrating the Rotliegend reservoir, almost half have been plugged. Formation water with high salt contents inducing corrosion and Zechstein salt creep were identified as major processes affecting well integrity.

3.1.1 Well Standard

Wells are constructed to provide a safe connection between the surface and the geologic reservoirs in the underground. To achieve this, steel pipes – called casings – are run into the hole and cemented in place by circulating well cement down the hole, around the shoe and into the annulus between rock and casing. For optimal cement placement the borehole is cleaned and the casing is centralized inside the borehole (Fig. 3.1a) before starting the cementation. The quality of the cementation is evaluated by pressure tests and/or sonic borehole measurements, called bond logs, over the full length of the cementation.

If a well is no longer needed, it is plugged. The necessary plugging operations are governed by regulations of the mining authorities. They require a bottom cementation to shut off the formation and further plugs above it to provide additional barriers. The intervals between the cement plugs are usually filled with heavy weight mud. To minimize leakage risks, the plugs are usually placed across potential problem zones as, for example, at the top of the liner (Fig. 3.1b). If uncemented zones are detected, the casing is perforated and cement squeezed into the annulus behind the casing. At the top of the well, the casing is cut approximately 3 m below the surface and covered by a concrete or a steel plate.

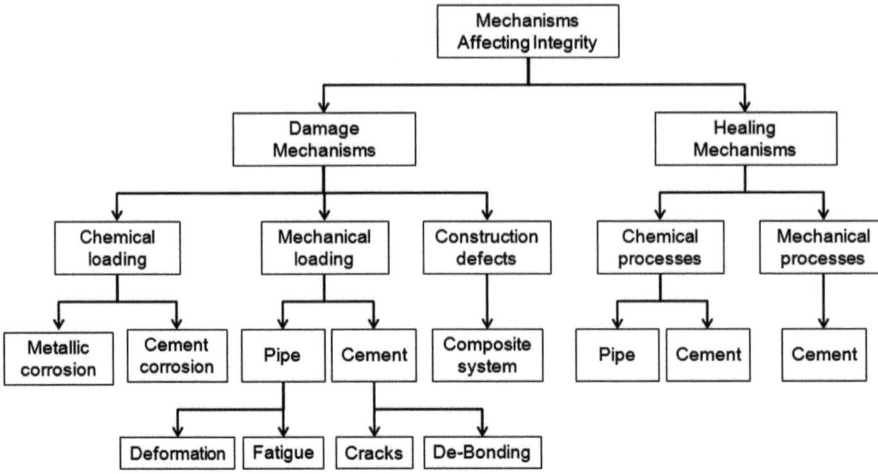

Fig. 3.2 Mechanisms affecting long-term well integrity

3.1.2 Technical Integrity

The term tightness of a system is only partly meaningful. Firstly, tightness cannot be measured but only be assessed within the measurement limits of the surveys used. Secondly, no technical system can be 100 % tight. The tightness of a well is consequently defined as a maximum acceptable leakage rate through the system or by means of other characteristics of well integrity. In integrity assessment of e.g. gas cavern wells an acceptable leakage rate of 50 kg/day was proposed as a criterion for technical integrity (Reinicke et al. 2009b; ITE 2007; SMRI 1996) in the 1990s, while the oil industry typically relies on pressure tests and a minimum interval of good quality cementation, when assessing zonal isolation capacity.

3.1.3 Mechanisms Affecting Technical Integrity

Figure 3.2 shows an attempt to classify the mechanisms affecting well integrity, distinguishing damage and healing mechanisms. Well integrity may be compromised as a result of poor well construction or abandonment, chemical degradation, or failures because of mechanical or thermal loading. As a result, CO_2 leakage pathways within the wellbore zone may occur, the wellbore zone being defined as the well and the immediate area around it. All mechanisms shown in Fig. 3.2 can be modelled.

3.1.3.1 Chemical Loading

The chemical processes induced by CO_2 influence all components in the wellbore zone. The metallic components like the casing are affected by electrochemical reactions leading to general weight loss and consequent failure. Wellbore cements are affected by CO_2 induced leaching processes, resulting in increased porosity and permeability, decrease in strength, and finally a loss of casing protection and zonal isolation. Publications addressing the carbonation of Portland cement systems are e.g. Bruckdorfer (1986), Krilov et al. (2000), Barlet-Gouédard et al. (2009) and Teodoriu et al. (2010).

Besides the dissolution processes impairing the cement stability, precipitation processes, resulting in decreased porosity and permeability and increased density and compressive strength, have to be considered as well. Carbonation does not only mean material removal but also re-crystallization, crystallization and related effects, respectively. Similar scientific observations with regard to salt cements (Reinicke et al. 2011b) are discussed below.

3.1.3.2 Mechanical–Thermal Loading

Mechanical-thermal loading of wells is to a large extend related to production and can be caused by high external pressure, cyclic loading by internal pressure resulting from intermittent production (ballooning/de-ballooning), declining internal pressure as a consequence of gas production, and tensile/compression strain induced by reservoir compaction or temperature variation during gas production. Relevant defects and damage mechanisms cause failure or even fatigue fracture within the composite system by exceeding the limits of material performance. As a result, damages can be classified in: failure by exceeding the limits of material performance (casing fracture, casing leak, casing collapse, cement cracking) or fatigue within the (casing-cement) composite system. In addition cyclic loading during the life of the well (long-term fatigue) must be considered for such wells (Teodoriu et al. 2010).

The mechanical self-healing phenomenon of wells penetrating natural salt formations is well known in drilling and, at the actual state of the art, ascribed to the plastic behaviour of salt. This salt property is considered as an essential criterion for suitability and safety of underground gas storages (Reinicke and Fichter 2009). Furthermore self-healing effects based on the plastic behaviour of clay are also widely known and applied, dependent on the geologic situation, in order to plug wells by building a second natural barrier (Williams et al. 2009).

3.1.3.3 Construction Deficits

In many cases poor well integrity results from construction deficits. Main causes for casing defects are wrong material or coupling selection. Frequent cement deficits are insufficient bonding of the cement to the casing and/or rock, development of "micro annuli", or development of channels in the annulus. Main causes for cementation defects are insufficient well and casing preparation, insufficient centralization, inappropriate cement recipes, cement circulation loss, or inflow after cementation.

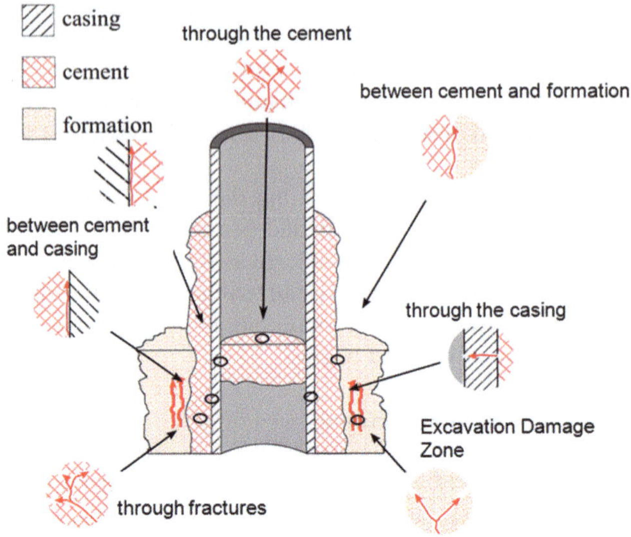

Fig. 3.3 Potential leakage pathways within the wellbore zone (Reinicke and Fichter 2010a)

3.1.3.4 Potential Leakage Pathways

Within the wellbore system, leakage can occur along a number of pathways through the wellbore system. These pathways result from the impacts on the system mentioned above. The system, consisting of the sequence of cement plugs and mud inside the casing for plugged wells, the composite system casing-cement-rock, and the pathways for leakages are shown in Fig. 3.3.

3.1.4 Evaluation of Well Integrity of Accessible Wells

Reliable evaluations, that CO_2 affected wells are tight and remain tight, require first of all adequate well information. For accessible wells, available well information may be supplemented by the results of additional wellbore surveys. For these wells, measurement strategies were developed to reliably assess their actual conditions. For abandoned wells, the available information may not be supplemented unless the wells are re-opened for surveying and/or re-plugging. For these wells, applicable assessment methods for the evaluation of technical well integrity were studied and applied.

3.1.4.1 Practices in the Gas Cavern Industry

In the gas cavern industry in Germany, well integrity is investigated by a pressure test on the last casing shoe, a so called leak-off test. The result of such a test, typically carried out with nitrogen, is clear evidence of well integrity, since the storage volume is below the casing shoe. In the 1990s Crotogino in SMRI (1996) suggested 50 kg/day as a minimum detectable leakage rate in such a test and 150 kg/day as a maximum admissible leakage rate. Since then measuring technology has advanced and values as to what constitutes a "significant leak" have decreased (Berest et al. 2002).

3.1.4.2 Practices in the Oil and Gas Industry

Leak-off tests are also carried out in the oil and gas industry in the course of well construction. Other than in the cavern industry, these leak-off tests, however, do not provide evidence for the integrity of the well in the direction out of the storage complex, since the reservoir volume, to which the well is to connect, is not below but above the casing shoe. Behind casing annular isolation is therefore verified indirectly via a minimum distance of contiguous good cementation across tight formations. Pressure tightness of casing and tubing is usually established in pressure tests. During these tests, typically carried out with water or gas, the volume to be tested for tightness is closed and pressured up as in the case of the recurring inspections of production tubular in thermal recovery wells (Reinicke and Fichter 2010b).

3.1.4.3 Strategies for the Integrity Assessment of Accessible Wells Affected by CO_2

The assessment of well integrity requires evidence that there is no leak in the downhole system and there is no significant fluid movement from the storage into an underground source of drinking water. There are numerous techniques to provide this evidence. Possible surveys to assess the conditions of casing and cement and their reliability in excluding e.g. leakage pathways through the cement have been discussed in Reinicke and Fichter (2010a), (Table 3.1).

The proposed strategy to determine the actual well conditions is to run a logging programme consisting of at least a Cement Bond Log (CBL)/Isolation Scanner to assess cementation quality and a Multifinger Caliper Log to assess the casing conditions. For the potential project in question, the logs should be run from the perforation area upward to cover the cap rock (here Zechstein salt). Depending on the results of the Caliper Log, the measurement should be extended to cover a longer section of the casing, for example, all the way to cover higher barrier formations. The use of an electromagnetic wall thickness measurement should be taken into consideration in case of more complex (overlapping) casing designs (Reinicke and Fichter 2010a).

To verify that there are no significant leaks, it is recommended to carry out a casing pressure test (annular pressure test) with water or nitrogen depending on the condition of the casing. If the result is negative, the test should be repeated to localize the leakage(s). Depending on the result of the pressure test, further testing may be required, e.g. temperature surveys, noise logging, radioactive tracer testing,

Table 3.1 Measurement methods (Reinicke and Fichter 2010b)

System	Damage	Survey
Casings	Corrosion internal	Caliper survey, isolation scanner
	Corrosion external	Isolation Scanner, electromagnetic wall thickness survey
	Weight loss	Isolation scanner, electromagnetic wall thickness survey
	Deformation	Caliper survey, isolation scanner, electromagnetic wall thickness survey
	Leakage	Caliper survey, isolation scanner
Cement	Annulus content	CBL, isolation scanner
	Micro annuli	CBL, isolation scanner
	Channels	Isolation scanner
	Cracks	(Isolation scanner)

and Oxygen Activation Log. The detection of a leak requires a successful repair before CO_2 operations are started.

Provided, the system has been verified to be free of significant leakages, attempts should be made to exclude significant fluid movements by use of cementation records. While the determination of a CBL derived bond index may suffice, it is recommended to base the conclusions on logging tools, which can identify and characterize local defects, rather than to provide only an average measure for the cement bond. If the bond index is used as criterion to declare behind casing annular isolation out of the storage complex, a minimum distance of 30 m of contiguous good cementation in the cap rock area with a bond index greater than 0.8 should be used in line with Hendricks (2009). This distance is significantly longer than what is used in the oil industry, which typically requires less than 10 m depending on the size of the cemented pipe. If 30 m of good cementation are not verifiable in the area of the cap rock, other measurements should be taken into consideration, for example temperature surveys, noise logging, radioactive tracer testing, and Oxygen Activation Log.

The behind casing pressure test (Reinicke and Fichter 2010b) is the last resort to investigate fluid movements behind pipe. The test will provide the desired evidence; it does however require milling (Abdel-Haq 2006) or perforating (Crow et al. 2010) the casing, potentially weakening a pressure tight and leakage free system.

3.1.5 Evaluation of Well Integrity of Plugged Wells and Predictions of Long-Term Integrity Development

The well database of the "Landesamt für Bergbau, Energie und Geologie (LBEG)" in the "Geozentrum Hannover" contains information on more than 30,000 wells, the majority drilled in the North German Basin. Almost 90 % of these wells are plugged and no longer accessible to assess their integrity. Also the area of interest for the CLEAN project in the Altmark contains plugged wells which have penetrated the cap rock of the potential storage horizon. The integrity of these wells, which may be affected by injected CO_2, needs to be verified prior to the commencement of an injection. The site-specific analysis of the behaviour of plugged and abandoned (p&a) wells to demonstrate their sealing capacity and quantify the risk of leakage in the long term is given in the following.

3.1.5.1 Strategies for the Integrity Assessment of p&a Wells Affected by CO_2 and Predictions of Long-Term Integrity Development

The evaluation of the technical integrity of plugged wells (Fig. 3.1b) is a challenge. The available well information may in many cases not be sufficient for a direct integrity assessment, but the information cannot be supplemented unless the wells are re-opened and surveyed. Evaluations require predictive methods based on available geologic and well information and in addition on a comprehensive understanding of the relevant THMC processes in the subsurface which affect well integrity. The methods are used, in addition, to predict the long-term development of technical well integrity under the influence of CO_2. These tasks are very similar to the integrity evaluation of, for instance, repositories, for which evaluation techniques have been developed over the years (Mazurek 2003) and in the meantime also applied to the sequestration of CO_2 by numerous investigators.

3.1.5.2 Methodology

The analysis comprises the three main steps characterization, scenario development, and consequence analysis. Characterization is achieved by a combination of Features, Events and Processes (FEPs). Features are static factors and parameters describing the site. Events are future occurrences, future changes to features and future alterations of processes. Processes are developments of the current and future aspects of the site. The interactions between the FEPs, that relate the intrinsic evolution of the system, are represented in an interaction matrix and a cause-effect diagram (Condor and Asghari 2009). The representation in an interaction matrix requires that the system, i.e. the well, is partitioned top down into constituent components, which are represented as the diagonal elements. Processes corresponding to interactions between the constituent components are recorded in the off-diagonal elements. The convention for illustrating the influence between components is that an off-diagonal element ij implies the influence of constituent component ii on constituent component jj if $i < j$ and vice versa if $i > j$. The representations provide an excellent basis for the definition of scenarios. Scenarios describe possible future developments of the system under consideration which in this case is the wellbore zone of abandoned wells, affected by the injected CO_2. The developed scenarios are evaluated in the last step, the consequence analysis. For the evaluation, numerical models are used generally. They describe the processes and events considered in a scenario in an abstract form.

3.1.5.3 Site Specific Analysis

The analysis addresses integrity of abandoned wells and long-term leakage risk in the wellbore zone, defined as the well and the immediate vicinity. Cap rock and overburden, abandonment elements, composite system, reservoir, reservoir fluids, and injected CO_2 were defined as main components of the system wellbore zone. To characterize the wellbore zone, the Quintessa FEPs database (Quintessa 2011) was chosen as a starting point, modified and supplemented to generate a case-specific wellbore zone FEPs database. The result consists of 56 FEPs. In a first approach, the interactions were described by the FEPs contained in the data base, generated in several expert sessions. For better conceiving, the entries in the matrix elements were then reduced to reflect the essential processes. Evaluations to determine their significance for risk of leakage followed. The result is shown in Table 3.2.

The evaluated interaction matrix and further representations, e.g. in a cause-effect diagram, provided a good basis for the definition of scenarios and the key processes that should be modelled in the scenarios. The general description of FEPs as well as relevant information and literature are documented in Reinicke et al. (2011a) and Reinicke and Weichmann (2011).

In a first attempt, two scenarios for the leakage of wells were distinguished:

- Leakage to the atmosphere and
- Leakage to overlying aquifers.

3.1.5.4 Model Development

The scenarios are being evaluated by using simulation models. For the simulation, the radial models shown in Fig. 3.4 are employed, one to simulate leakage to the atmosphere and the other to simulate leakage to an overlying aquifer. In steps of increasing complexity, single-phase and two-phase flow simulations were carried out, excluding and including mechanical and chemical effects.

For the single-phase and two-phase flow simulations Schlumberger's numerical simulation software packages E100 (Black Oil Simulator) and E300 (Compositional Simulator) were used. For the two-phase flow simulation several relative permeability models were evaluated to identify a model best suited to describe the flow in neat and fractured cement. For the simulations including mechanical and chemical processes more advanced simulators are planned to be used in the future.

In the simulations, the reservoir is modelled as a closed tank containing 100,000 t of CO_2. Reservoir properties are assigned to the reservoir part of the model, representative of the reservoir of interest, i.e. a depth of 3,000 m, a temperature of 125 °C, a porosity of 15 %, an initial water saturation of 20 %, and a permeability of 100 mD (mD = milli Darcy, 1 mD = $1 \cdot 10^{-15}$ m^2). Two values were assumed for the reservoir pressure: The first pressure of 5.0 MPa (0.1 MPa = 1 bar) reflects the current pressure while the second pressure of 35.0 MPa resembles initial conditions. For this model potential leakages were

Table 3.2 Interaction matrix,*T* thermal, *H* hydraulic, *M* mechanical, *C* chemical, *P* pressure (After Reinicke et al. 2011a)

	1	2	3	4	5	6
1	**Reservoir**	[1,2] a) Decompaction b) TM-loading	[1,3] a) Decompaction b) TM-loading	[1,4] a) Decompaction b) TM-loading	[1,5] a) C-equilibrium b) Solution, Precipitation	[1,6] a) C-equilibrium b) Solution, Precipitation
2	[2,1] a) Containment b) Decompaction	**Cap Rock**	[2,3] a) (Creep) Loading	[2,4] a) (Creep) Loading	[2,5] a) Containment	[2,6] a) Containment
3	[3,1] a) Loading resistance	[3,2] a) Loading resistance b) Salt partial solution	**Composite System**	[3,4] a) Abandonment conditions b) TM-loading	[3,5] a) Sorption, Filtering	[3,6] a) Sorption, Filtering
4	[4,1] a) Sealing b) Abandonment conditions	[4,2] a) Sealing	[4,3] a) TM-loading b) Sealing	**Abandonment Elements**	[4,5] a) Sorption, Filtering	[4,6] a) Sorption, Filtering
5	[5,1] a) Solution, Precipitation	[5,2] a) P-loading	[5,3] a) Defects b) P-loading c) C-loading d) Solution, Precipitation e) Fluid advection f) Gas buoyancy	[5,4] a) Defects b) P-loading c) C-loading d) Solution, Precipitation e) Fluid advection f) Gas buoyancy	**Reservoir Fluids**	[5,6] a) C-equilibrium b) Solution, Precipitation
6	[6,1] a) P-loading b) Solution, Precipitation c) CO_2 advection d) Water transport	[6,2] a) P-loading b) Containment c) CO_2 advection d) CO_2 diffusion	[6,3] a) Defects b) P-loading c) C-loading d) CO_2 advection e) CO_2 buoyancy f) CO_2 diffusion	[6,4] a) Defects b) P-loading c) C-loading d) CO_2 advection e) CO_2 buoyancy f) CO_2 diffusion	[6,5] a) C-equilibrium b) Solution, Precipitation c) Fluid displacement	**Injected CO_2**

Risk: 0 = zero, 1 = very low, 2 = low, 3 = medium, 4 = high

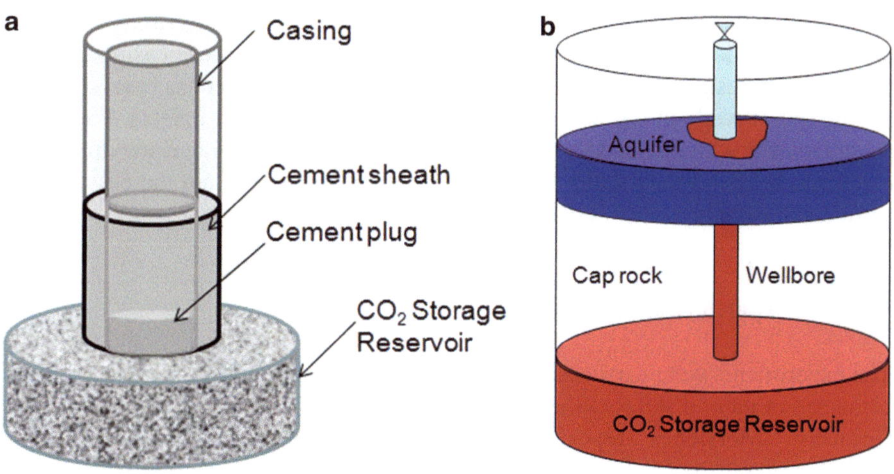

Fig. 3.4 CO_2 leakage to the atmosphere (*left*) and to an overlying aquifer (*right*)

simulated for 1,000 years assuming three different values for the cement permeability: a permeability value representing good cement, poor cement, and highly fractured cement or cement with channels.

Tables 3.3 and 3.4 show the results for a "worst case" of single phase flow of gaseous CO_2 leaking to the atmosphere from a reservoir with a pressure of 35 MPa and the case representing the pilot project with CO_2 leaking to the atmosphere under two phase flow conditions from a reservoir with a pressure of 5 MPa.

Table 3.3 Results for single phase flow to the atmosphere for a reservoir with 35.0 MPa pressure

Cement quality	Permeability (mD)	Leakage volume fraction of IGIP[a] (%)	Pressure drop (%)
Good cement	0.01	0.001	0.03
Poor cement	10	1.14	1.6
Fractured cement	1,000	54.5	57.7

[a]*IGIP* initial gas in place

Table 3.4 Results for two phase flow for a reservoir with 5.0 MPa pressure

Cement quality	Permeability (md)	Leakage volume fraction of IGIP (%)	Pressure drop (%)
Good cement	0.01	0	0.02
Poor cement	10	0.003	0.2
Fractured cement	1,000	2.82	4

3.1.6 Salt and Compaction Influence on Well Integrity

According to the FEPs analysis of the previous chapter, mechanical and chemical processes are important when assessing well integrity and the long-term development of the well zone. The literature on CO_2 sequestration is full of information on processes causing defects. Little is found on processes preventing defects to occur or healing defects. These processes are equally important in describing the long-term integrity development of wells. Salt creep is known to cause casing collapse for wells not designed to withstand plastic salt pressures, which has been observed for the initial wells in the area of interest, but it can also heal defects. Formation waters loaded with salt and CO_2 can corrode steels, but precipitates may passivate metal surfaces and potentially heal defects, which would explain the cessation of sustained casing pressures, observed for some wells.

3.1.6.1 Experimental Tests

The experiments were carried out on full-scale specimens, consisting of steel pipes cemented in natural salt rocks mined from Zechstein in different mines in Northern Germany (Morsleben repository, Bernburg and Siegfried-Giesen). Hollow cylinders of salt rock were produced by centrally coring the salt cylinders (outer diameter OD = 280 mm, inner diameter ID = 140 mm, length L = 620–650 mm). In the borings, 20 mm N80-steel casings (ID = 81.6 mm, OD = 101.6 mm) were inserted and the annuli cemented with salt cement based on Dyckerhoff Well Cement black label, API Class G, high sulphate resistant. The salt saturated mixing water was applied in line with field practice to avoid dissolution of the natural salt rocks as a consequence of water involved in the binding process. During cementing, artificial capillaries were created by inserting wires (OD = 1.9 mm), which were regularly moved in order to avoid binding with the cement.

Fig. 3.5 Full-scale specimen types of the composite system salt-cement-casing (Reinicke et al. 2011b) with type 1 – position of defect within cement (*left*), type 2 – position of defect at contact cement-casing (*centre*), and type 3 – position of defect at contact salt-cement (*right*)

Three positions were chosen for these artificial defects (Fig. 3.5): inside the cement sheath (type 1), at the cement-casing interface (type 2), and at the cement-salt interface (type 3). After complete binding and setting for 28 days, the wires were removed. The three specimens were then tested under in-situ conditions in triaxial cells (Zhang et al. 2011).

Self-healing was observed for all samples tested. In the experiments the CO_2 permeability decreased within 50–60 days from initially approximately 10^{-13} m^2 to values of 10^{-18}–10^{-22} m^2. For the defect at the salt-cement interface (type 3), this was in line with the expectation that salt creep would heal the defect. The rapid decrease in permeability for the specimens with artificial defects at the cement-casing interface (type 2) and within the cement sheath (type 1) was a surprise.

In order to examine the causes for the permeability decrease of the artificial capillaries, the specimens were cut open and inspected. In the capillaries secondary healing was visible to the unaided eye. In the case of the capillary within a cement sheath, the healing was the result of a green-yellow crystalline deposit. As established by means of X-Ray Defraction (XRD) analysis the deposit consisted of salt minerals, i.e. halite (NaCl), potassium halite ($K_{0.4}Na_{0.6}Cl$), calcite ($CaCO_3$) and traces of a clinker phase (larnite, α-Ca_2SiO_4).

3.1.6.2 Self-Healing Phenomena

Self-healing phenomena in cementitious materials are well known from the construction industry and related sciences and have been frequently documented (Jacobsen et al. 1998; Schlangen and de Rooij 2005; Barlet-Gouédard et al. 2009). Especially the carbonation effect of hydraulic binders like cement and concrete in interaction with CO_2 containing water is the main phenomenon, which causes densification as well as an increase in hardness of the binder due to a rearrangement of material, apart from corrosion of metal materials and solution effects. Concerning the occurrence of such phenomena in the subsurface of wells, it has been found that Portland cement systems show similar effects (Bachu and

Bennion 2009). The existence of carbonate and salt minerals in Portland cement based systems has been described and interpreted by Carey et al. (2007) and Crow et al. (2010). In contrast to these findings for fresh water based cement systems, the literature does not provide information on the behaviour and reaction of salt cements under the influence of CO_2. Salt cements are typically used when drilling through salt formations, as for example in the wells in the area of interest. Results for this type of cement have been published for the first time within the context of the CLEAN research work presented here.

The experimental results obtained suggest, that defects heal within a few weeks under constant flow of dry CO_2, with permeability decreases to values below 10^{-20} m^2, values which correspond to the permeability of the surrounding virgin rock salt. However, longer testing periods are considered a necessary requirement for future experimental work.

3.1.7 Concept for Well Integrity Assessment

The current standard of constructing and abandoning deep wells in Germany is high. Nevertheless risks of leakages through wells remain and have to be further investigated. Risk mitigation requires the assessment of current well integrity and its future development. Accessible wells pose no problem. Their current integrity can be assessed through measurement surveys and tests, and integrity can be monitored to intervene when leaks are detected.

Plugged wells are no longer accessible for direct measurements and therefore cannot be surveyed to assess their current status. Integrity evaluation of plugged wells requires predictive methods based on the available geological and well information and on a comprehensive understanding of the relevant THMC processes during construction, operation and after abandonment, which affect well integrity. If the available information is insufficient for a conclusive evaluation, integrity development must be monitored or additional information must be obtained. Suitable scientific methodologies have originally been developed for the evaluation of repositories and are now widely applied to assess the suitability of CO_2 storages. Application of the methodology to the area of interest in the CLEAN project suggests that chemical and mechanical processes are important for the assessment of well integrity.

Tri-axial experiments under in-situ conditions on full-scale samples of casing cemented in rock salt provide evidence, that chemical and mechanical processes not only pose threats to well integrity but also provide mechanisms of self-healing. In particular, it was shown, that self-healing in the composite system casing-cement-rock salt must not only be ascribed to the plastic behaviour of salt but also to self-healing mechanisms resulting from the interaction of salt, cement and casing with dry or wet CO_2. While these observations are encouraging, further work is required to fully understand and quantitatively model the processes leading to the observed healing of defects.

To assess well integrity and quantify leakage risks requires reliable information on the wellbore zone, experience in data integration and evaluation, as well as knowledge of all processes in the wellbore zone which affect integrity. In evaluating well integrity and leakage risk, modelling should not be restricted to processes leading to damages as observed in the laboratory for relatively short time frames. A realistic integrity/risk assessment has to consider the wellbore zone as a whole, taking into account also self-healing processes, resulting from the interactions of the CO_2 with the components of the wellbore zone and the reservoir fluids.

3.1.8 Influence of Metal Corrosion on Well Integrity

Whenever CO_2 is accompanied by water carbonic acid is produced which is corrosive to metallic materials, specifically low alloy steels as used for pipelines and downhole tubings and casings. This is well known in oil and gas production and the corrosion mechanisms have been described for CO_2 partial pressures up to 2 MPa, including effects of by-products like H_2S, low molecular weight carbonic acids (e.g. acetic acid) and oxygen in the ppm concentration range. However, for systems with highly concentrated CO_2 compressed to high pressures and handled under liquid or supercritical conditions much less information is available on the performance of steel construction materials under such conditions up to elevated temperatures.

To quantify the long-term integrity development of wells and other equipment for CO_2 injection in the Altmark field experimental investigations were performed within the CLEAN project (Schmitt et al. 2011a–h, Schmitt et al. 2012). The test conditions considered worst case scenarios: on the environmental side supercritical (sc) CO_2, presence of water, flow (4 m/s), temperatures from 50 °C to 130 °C; on the materials side low alloy pipeline steel X65, low alloy tubing steel 38Mn5/C75, martensitic 13Cr steel (X20Cr13; UNS 42000;1.4021), austenitic-ferritic (duplex) steel (X2CrNiMoN22-5-3; UNS S31803; 1.4462) and high alloy austenitic stainless steel (X1NiCrMoCu 25-20-5; UNS N 08904;1.4539; alloy 904 L).

In the absence of a separate water phase $scCO_2$ is not corrosive for all alloys tested. However, water mist, in equilibrium with $scCO_2$ causes corrosion at low alloy (carbon) steels, specifically pitting at sites of droplet impacts. High alloy Cr and CrNi steels proved resistance even up to 130 °C. Water in equilibrium with a $scCO_2$ phase is very corrosive for carbon steels yielding intense mesa-type localized attack (5–15 mm/year at 4 m/s) in the whole temperature range of 50–130 °C. Under such conditions 13Cr steel is not sufficiently resistant (corrosion rates 0.3–0.8 mm/year). Even the austenic-ferritic (duplex) and the austenitic CrNi steel (alloy 904 L) surpass the target line of 0.1 mm/year at 110 °C. The presence of small concentrations of salt (1,000 ppm chloride as NaCl) in the water phase increases slightly the intensity of localized attack at carbon steel and 13Cr steel, the CrNi steels remain unaffected. It is expected that higher salt concentrations increase the intensity of localized attack. Corrosion inhibitors can reduce corrosion rates in $scCO_2$ systems significantly. While uneconomically high inhibitor concentrations are needed to reduce the corrosion of carbon steels below 1 mm/year, high alloy steels need only 100 ppm of typical CO_2 corrosion inhibitors to keep penetration rates below 10 μm/year (Schmitt et al. 2011 a-h).

Only small oxygen concentrations in the $scCO_2$ phase (1,000 ppm down to only 2 ppm O_2 based on the mass of CO_2) increase the intensity of materials attack at carbon steels up to 200 mm/year. Under such conditions even 13Cr steel experiences corrosion rates up to 0.3 mm/s. Surprisingly, higher oxygen concentrations in the $scCO_2$ phase in the order of 20,000–60,000 mass ppm seem to "passivate" carbon steels yielding corrosion rates of only 1–2 mm/year. All Cr containing steels remain passive under these conditions (Schmitt et al. 2011 a-h). This oxygen effect and threshold values for H_2S and CO impurities in the CO_2 need more detailed investigation.

The corrosiveness of aqueous phases in equilibrium with $scCO_2$ is comparable with the aggressiveness of non-supercritical CO_2 corrosion systems with high CO_2 partial pressures. The corrosiveness depends on the carbonic acid concentration and the growth kinetics of corrosion product films at given temperatures and pressures. At temperatures above 100 °C iron oxides can become part of the corrosion product films on carbon steel due to iron oxidation by water in the gas and/or supercritical phase. From all CO_2 corrosion prediction models presently available only the thermodynamics based OLI software (http://www.olisystems.com/) yields reliable trend information on the corrosion resistance of carbon steels and 13Cr steel in $scCO_2$/water systems (Schmitt et al. 2012)

Based on the experimental results recommendations can be given for the CO_2 quality (final specification still to be verified) and the materials selection for CO_2 injection wells and other equipment in the Altmark field. First of all, the CO_2 to be transported and injected needs to be dried to a dew point of 25–30 °C before pipeline transport and injection. Based on the planned injection conditions (injection of gaseous CO_2 at 10–40 °C with flow rates $<$15 m/s) and considering the relative short duration of the test injections, the completion design of injection wells can be kept simple. The wells shall be made from 13Cr steel. This assures well integrity also under conditions when for a limited time period the dew point is higher than specified. The completion of injection wells should include packer, subsurface safety valve, nipples for plugs and tools, no flow couplings, no downhole shut-off valve, no sliding side door, no expansion joint, no side pocket mandrel, no inhibition. To prevent back flow of wet gas during longer down times plugs shall be set. At the production wells in the Altmark which are presently completed with carbon steel without packers (no free formation water has been observed so far), permanent CO_2 monitoring should be performed. At arrival of CO_2 from CO_2 injection re-completion should be made from 13Cr steel including packers. This holds as well for observation wells should an increase of CO_2 concentration be observed in the gas.

3.2 Monitoring and Intervention

CO_2 wellbores are supervised using effective methods and technologies of monitoring and intervention during the injection operation phase (30–50 years) in order to guarantee the mining safety of CO_2 underground storage. After injection, when the pressure of the CO_2 reservoir reached its final level, all well bores need to be abandoned for the long and safe post operational phase (hundreds to thousands of years). Figure 3.6 gives an overview of the methods relevant in practice for monitoring and intervention. Some of these methods were adapted to the CO_2 case based on experiences of natural gas storage; the others were specifically developed for CO_2 wells (Fig. 3.6).

The monitoring methods are focused on the mechanical stability and the gas tightness of the wellbore which is given by the system casing – cemented annulus – cap rock as the most critical part of the well. Further, monitoring controls the CO_2 flow processes mainly in the tubing. The borehole simulator BOCO (DBI Gastechnologisches Institut gGmbH) was specifically developed in order to simulate the flow processes thermodynamically of CO_2 between well head and bottomhole in terms of operational safety. An eruption of CO_2 from a well represents the most critical incident to be avoided in any case. However, the consequences of such event are controllable as shown by the presented simulation results.

The methods of intervention are focused on the leaking casing or tubing and on the cement repair during the injection phase using technologies of secondary sealing. Another field of application is the long-term abandonment of CO_2 wells by CO_2 resistant natural materials exhibiting plastic and self-healing properties.

3.2.1 Monitoring and Simulation Methods

3.2.1.1 Well Logging

Geophysical well logging is an indirect monitoring method for the in-situ state of casing and cement. The most effective logging methods are to be selected for the best measurement results in order to conclude on the sealing properties of the system casing – cement – rock. Figure 3.7 lists the methods of well logging used in the years 1975–2000 and those to be employed during the workover of the wells before the CO_2 injection.

The former well logging results can be interpreted as prove for the gas tightness of the system casing-cement-rock, which hence does not require intervention.

3.2.1.2 Pressure Survey of the Well Annuli

Measurement of the well head pressure in the cemented annuli behind the casing is the best indication of any CO_2 migration through the cement. In the case of a gas tight system of casing and cement, the well head annulus pressure remains zero.

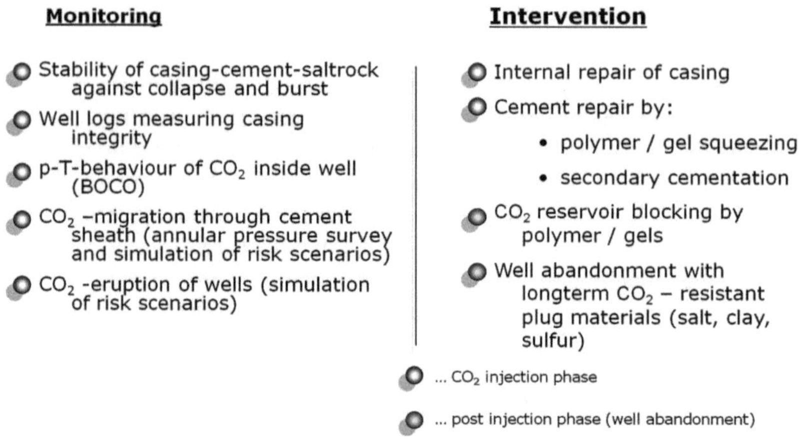

Fig. 3.6 Overview of effective methods for monitoring and intervention of CO_2 well bores

Casing and cement evaluation by well logging

a) Logs of 1975 – 2000

- Correlation log
- Cement log (ZL, cement bond)
- Multyfinger calibre (MFC, inner tube)
- Gamma ray and neutron-neutron (GR-NN)
- Special measurements

Conclusions

- Marginal losses in wall thickness during production
- Good cement binding, particularly in salt cap rock
- State of casing joints permits no clear interpretation

→ Old logs confirm integrity of the Aaz wells without necessary intervention

b) Logging planned until and during CO_2-injection

Process	Measuring interval	Information	Remarks
CBL	Accessible well	Cement bond, joints	Only fluid filled well
Multifinger calibre	Entire casing, tubing	Material loss, corrosion, casing position	Maximum number of "fingers"
Electromagnetic wall thickness meter (USIT)	Entire casing	Wall thickness, internal + external corrosion, condition of joints	Check Liner condition
Temperature	Accessible well	Overflow	Consider standstill time
Gamma	Accessible well	Lithology, depth	in combination
CCL	Entire casing	Joints position	in combination
GG/neutron	Accessible well	Overflow, cement, gas accumulation	

Fig. 3.7 Well logging methods for the CO_2 wells of the pilot field in 1975–2000 and for the future

In case of slight pressure build up, the accordingly low CO_2 migration rate can be accepted applying a periodic pressure release through the head valve. In case of higher rates, cement repair will be required. Figure 3.8 shows the simulation model for the migration of CO_2 through the cement sheath between well bottom and head.

Applying this model, various migration scenarios of pressure build up at the wellhead can be computed or – more relevant – the wellhead pressure development measured in the field enables conclusion on the state of cement. In Fig. 3.9, such scenarios are illustrated.

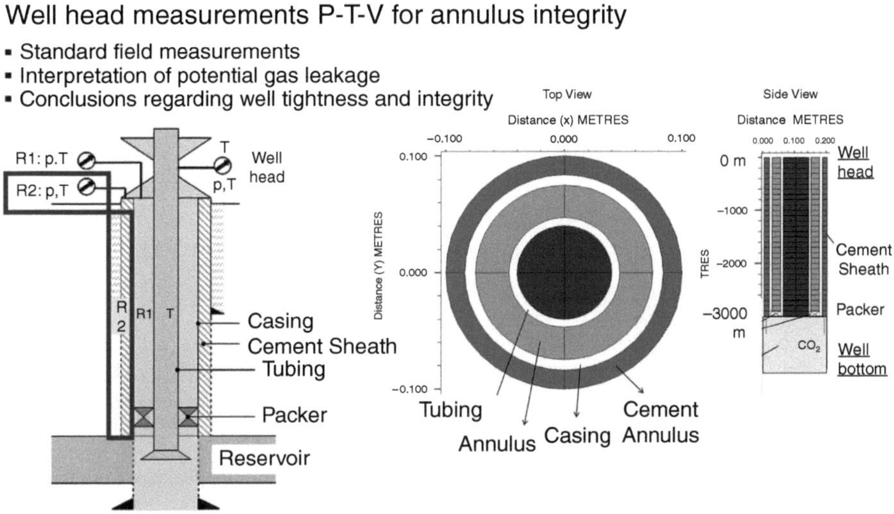

Fig. 3.8 Simulation model of CO_2 migration through the cemented well annulus

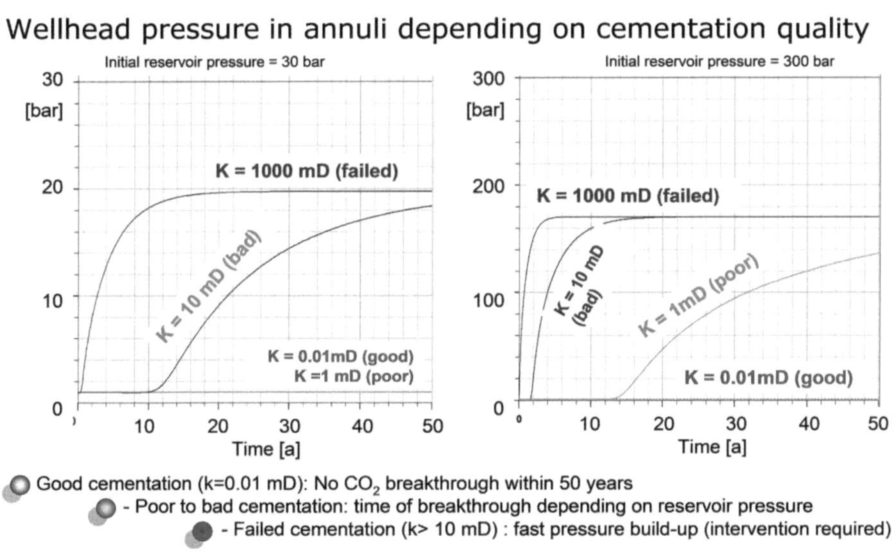

Fig. 3.9 Simulation of wellhead pressure development for various cementation qualities at 3 and 30 MPa initial reservoir pressure

At a low reservoir pressure of 3 MPa (30 bar) – as it is the case for the pilot field – a bad or failed cementation can be detected by annulus pressure build up with the conclusion of necessary intervention by cement repair. In the case of high reservoir pressure of 30 MPa (300 bar), only the good cementation guarantees acceptable closure of the system whereas poor and bad cementation require repair by polymer injection, the failed cementation by squeezing with powder cement or polymer gels.

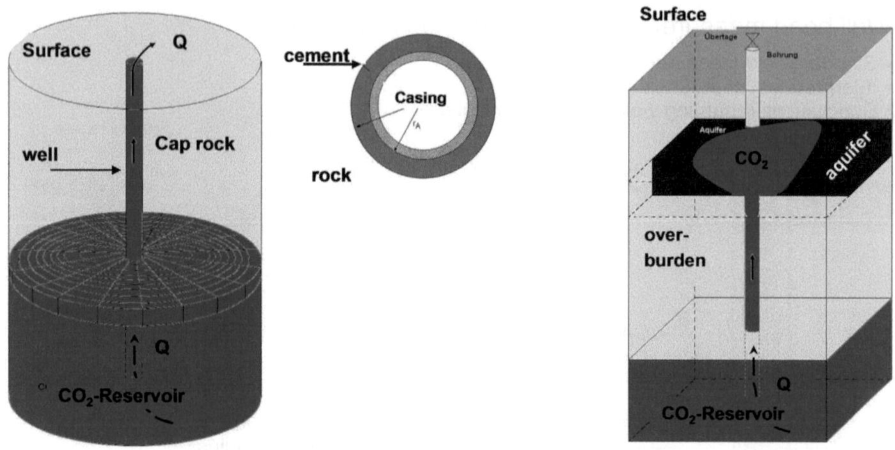

Fig. 3.10 ECLIPSE Model of vertical CO_2 migration – scenarios for direct leakage from the well into the atmosphere (*left*) and into an overlying aquifer (*right*)

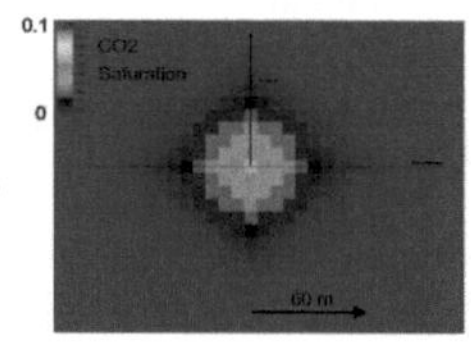

Cementation good | poor | failed

	k=10^{-2} mD, Q=$2*10^{-4}$ m³/d			k=1 mD, Q=$2*10^{-2}$ m³/d			k=10^{3} mD, Q=20 m³/d		
Time t [a]	V [m³]	p [bar]	R [m]	V [m³]	p [bar]	R [m]	V [10^6 m³]	p [bar]	R [m]
0	0	100	0	0	100	0	0	100	0
10^2	7,3	100,01	~ 0	730	100,08	~ 12	0,73	100,23	110
10^3	73,0	100,02	~ 10	7 305	100,19	~ 32	7,3	100,23	210
10^4	730,0	100,09	~ 25	73 067	100,23	~ 52	73,0	100,24	700
10^5	**7 304,0**	**100,21**	**60**	730 614	100,23	100	730,0	100,27	1700

Fig. 3.11 Simulation of CO_2 migration cases depending on cement quality and time (assumed reservoir pressure 30 MPa = 300 bar)

3.2.1.3 CO_2 Migration Scenarios

Coupling of the former model of cement annuli with a model of CO_2 accumulation above the reservoir, assuming either a direct migration into the atmosphere or – more realistic – a temporary capture of CO_2 within an overlying aquifer – a migration model as shown in Fig. 3.10 is set up for estimation of release and accumulation of CO_2 over time depending on cementation quality.

Some simulation results for various cementation qualities are given in Fig. 3.11 for the case of known cement permeabilities and a CO_2 reservoir pressure of 30 MPa. In case of good cementation (k = 10^{-2} mD), only low or negligible

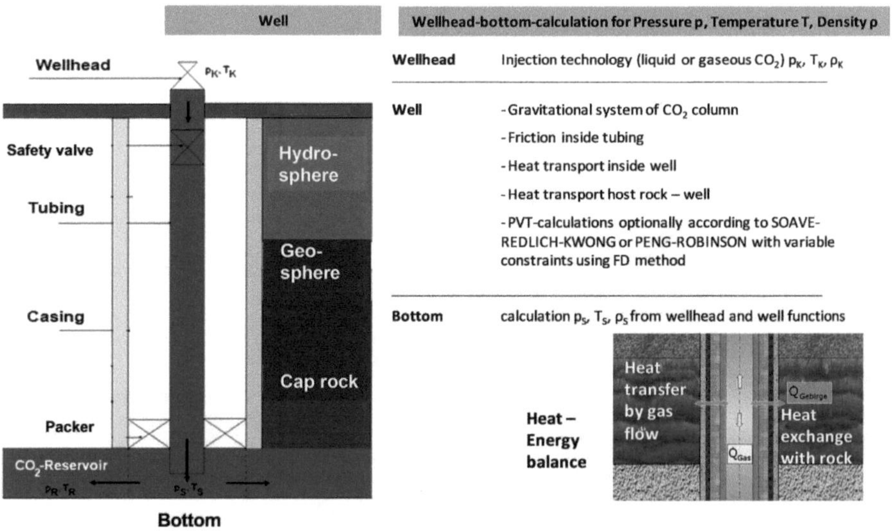

Fig. 3.12 BOCO thermodynamic well model

CO_2 migration rates are observed: After 10,000 years, 700 standard m^3 have migrated into the atmosphere or in case of the overlying aquifer form an accumulation of 25 m in radius in this layer. Such a small volume of CO_2 practically is not measurable. After 100,000 years, about 7,000 standard m^3 have migrated or accumulated with an extension radius of 60 m, accordingly. This still represents a negligible amount compared to the stored volume of 730 Mm^3 CO_2. However, in case of failed cementation ($k = 10^3$ mD), the situation is different: After 10,000 years, 10 % of the stored volume has migrated, and after 100,000 years of leakage, the reservoir practically is depleted.

3.2.1.4 Thermodynamical Well Model

The thermodynamics of CO_2 in the range of common pressures and temperatures between well head and bottom during injection into a depleted gas reservoir strongly differ from those of natural gas production. In particular, the phase transition gas–liquid requires consideration. In terms of operational and mining safety, this phase transition must be avoided inside the tubing in order to prevent pressure pulses. To compute the entire range of PVT-conditions in a well covering the heat transfer with the surrounding rock mass as well, the model BOCO was developed to control injection and/or production of various gases from a well (Fig. 3.12). The model was validated with data from an actual case of CO_2 injection and its reliability shown for injection of gaseous as well as liquid CO_2. However, the phase transfer between liquid and gas inside the well cannot be calculated so far. BOCO was used for the process simulation of CO_2 gas injection into a well of the pilot site.

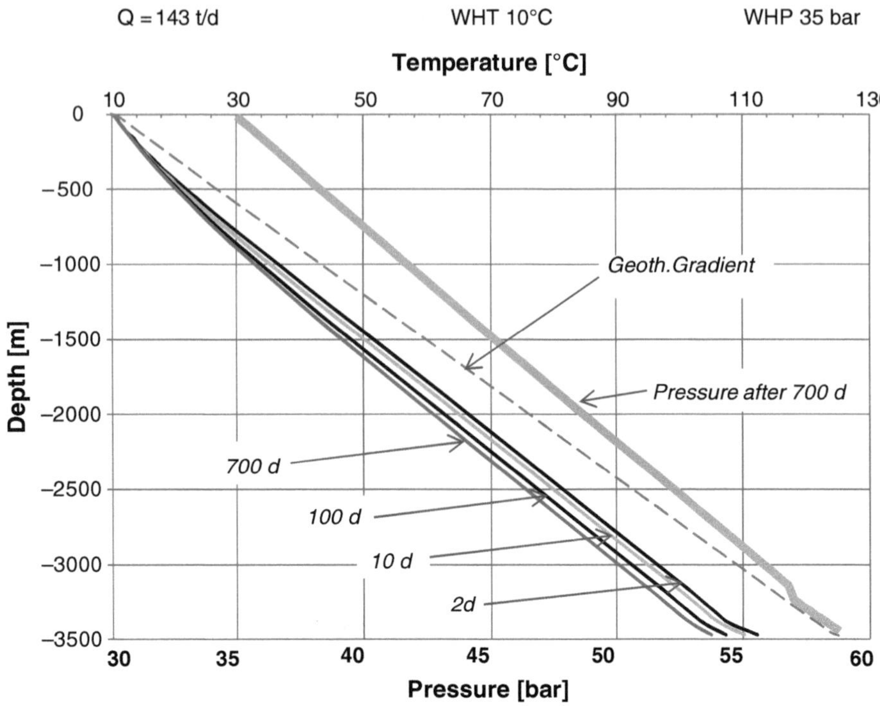

Fig. 3.13 Pressure–temperature cross section in the well of the pilot site, computed with BOCO

An example of a "cold" gas injection is given in Fig. 3.13. In this particular case, the conditions of the injected gas at the well head are WHT (well head temperature) = 10 °C and WHP (well head pressure) = 3.5 MPa in order to ensure no phase transition and to minimize the amount of energy required at the surface plant for re-heating of the liquid CO_2 (delivered by trucks in liquid state). The borehole temperature decreases compared to the geothermal gradient by 20 K as long as injection is continued. The state of CO_2 remains gaseous down to the bottom of the well. Considering a case of WHT = 20 °C and WHP = 5 MPa, CO_2 reaches the supercritical state below 1,500 m depth. Both cases, gaseous as well as supercritical CO_2 do not provoke the risk for the tightness of the system casing – cement – rock. In the described cases the CO_2 injection in the pilot well can be controlled.

3.2.1.5 CO_2 Eruption Scenarios

The most dangerous accident of a CO_2 well is a CO_2 eruption through the tubing or casing. Such an event – to be avoided in any case – can however not be excluded for an enhanced gas recovery or a storage project. Therefore, the free outburst of CO_2 into the atmosphere, which might occur until final closure of a well, requires an assessment of the safety around and at the well site.

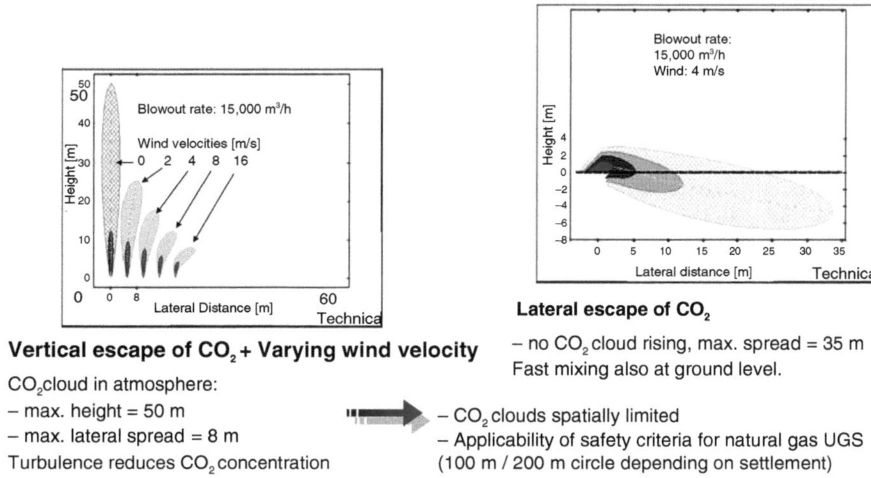

Fig. 3.14 CO_2 eruption scenarios modelled with the simulation tool TECJET

The simulation tool TECJET used in natural gas storage was applied in order to compute the effect of the CO_2 plume in the environment of the well head. The simulator was adapted with regard to the gas dynamics of CO_2. Figure 3.14 illustrates simulation results of a vertical and a lateral eruption. The main observation are a narrow shape of the CO_2 plume spreading into the atmosphere and the strong mixing between CO_2 and air in the case of wind turbulence. Within the general safety circles of 100 and 200 m around a well, requested by mining authorities for natural gas storage sites, CO_2 is diluted down to background values by the adjacent air and does not bear any threat for people outside these circles. Thus, despite the higher density of CO_2 compared to air, CO_2 behaves similarly to other gases above ground due to the eruptional pulse. The eruption from a well can be killed by conventional tools and methods of the gas and oil industry.

3.2.2 Intervention

3.2.2.1 Casing Repair

The typical reasons for a leaky casing or tubing are fissures in the pipe or non-sealed tool joints. The most appropriate technology of repair is patching. As shown in Fig. 3.15, a thin metal foil is pressed against the casing at the inner side of the leaking section of the pipe or joint in order to seal it. In case the patching method fails, the common heavy repair technology of casing/tubing exchange has to be applied.

3.2.2.2 Cement Repair with Polymer Gel Solution

Formerly, powder cement suspensions were commonly applied for the operation of cement repair. The squeezing technology is illustrated in Fig. 3.16 in principal.

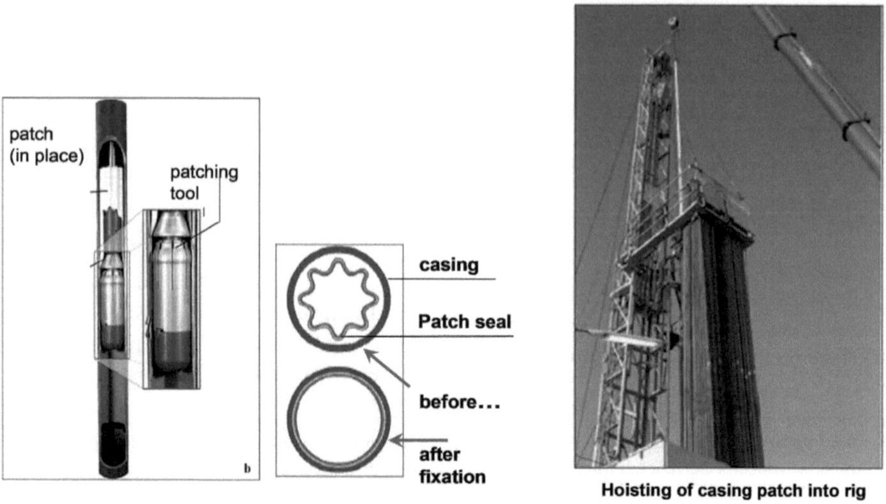

Fig. 3.15 Patch technology for internal sealing of casings or tubings

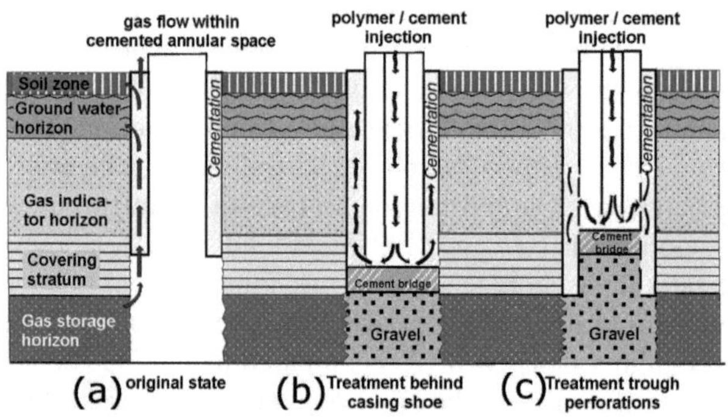

Technology:
- Injection of polymer solutions in to cement mantle around casing shoe (b)
- Radial-vertical polymer injection through casing perforation in cement mantle (c)

Experience:
- Laboratory tests on cement cores with polymer injections → reduction of permeability at range of 10^{-1} (K=4*10^{-1}mD / 10^{-16} m² →K=4*10^{-2} mD / 10^{-17} m²)

Usable polymers: 1) NaSi (sodium silicate-acid-solution) + CO_2
2) PAM (polyacrylamide-solution) + crosslinker

Fig. 3.16 Cement repair behind the casing by polymer or cement injection

However, squeezing with aqueous polymer solutions leads to better results because of the better infiltration of the cement rock in respect of the absence of a filter cake.

These solutions form a gel in fissures and pores of the cement body. Two base recipes for the polymer injection were developed as remarked in Fig. 3.16: Sodium silicates with CO_2 or polyacrylamide with cross linker. These polymer gels lead to impermeability of the cement.

The technology was successfully applied in leaky gas storage wells. In laboratory cement rock probes were furnished with vacancies (cracks) which afterwards were

Table 3.5 CO_2 resistant cements (after Nygaard 2010)

Name	Description
Pozzolanic Portland cements	The addition of Pozzolan can reduce permeability and minimizes chemical attack from some types of corrosive formation water
Microfine cements	Cements composed of very finely ground cements of either sulphate-resisting Portland cements. Average and maximum particle size is 4–6 and 15 μm, respectively. These cements penetrate relatively small fractures
Latex cements	Latex cements are blends of API Class A, G or H cements with the polymer latex added. The additive may protect the cement from chemical attacks, such as formation water containing carbonic acid
Portland-alumina/ calcium cements	Al-/Ca-hydrates reduce the Portland cement concentration for increased CO_2-resistance for use in acid gas wells

sealed by polymer injection. On the analogy of the real technology of polymer squeezing the gel (e.g. water glass) was pressed in the cracks by pressure injection. By the treatment with gel a permeability reduction of tenfold was reached and the permeability of a cracked cement core was reduced to $1.0 \cdot 10^{-2}$ mD. Hereafter the treated cement cores were stored in autoclaves under CO_2 atmosphere at 10.0 MPa and 50 °C to examine the resistance of the polymers with regard to supercritical CO_2.

3.2.2.3 Cement Repair with Cementitious Materials

The cement selected for a squeeze cementation in a well exposed to CO_2 has to be no less resistant to CO_2 corrosion than the cement in the cement sheath. In addition slurry design will depend on the technique to be used and the type of damage to be repaired. Table 3.5 provides an overview of available CO_2 resistant cements.

3.2.2.4 Restrictions of Cement Squeezing

The repair of leaks in the cementation behind the casing with cementitious material is restricted to defects large enough to accept the cement slurry. Small cracks with apertures less than approximately 15 μm are outside the application range of this technique. The microfine cements have an average particle size of around 5 μm. Particles of this size will bridge in fissures less than three times the size of the particles. The radial cracks with apertures of 5–10 μm as the result of cyclic loading, e.g. ballooning and de-ballooning, during the operational life of a well can therefore not be remedied by use of cementitious materials. Against this background it will be difficult to repair typically occurring micro-annuli of up to 300 μm.

There is a general consensus among experts that the most successful method is the low-pressure hesitation squeeze with a low fluid loss slurry and a packer or retainer as isolation tool. The high pressure squeeze technique is not recommended. Mixed results are reported on the success rate of squeeze operations. To mitigate this problem, best-practice guidelines are offered based on long field experience.

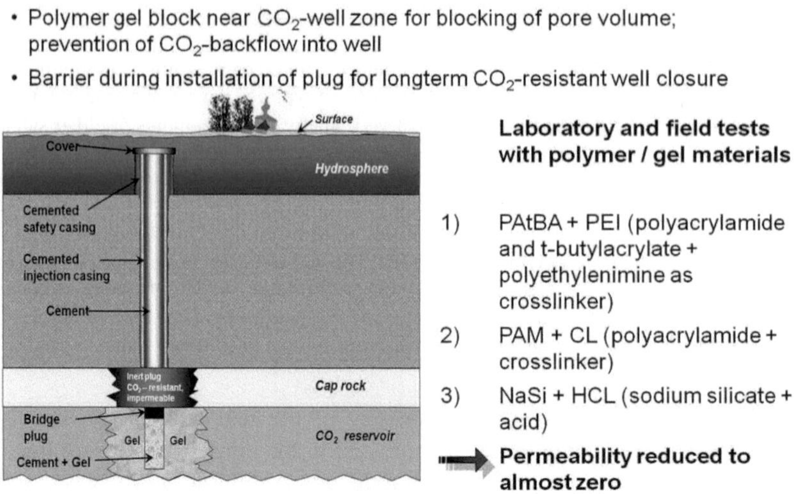

Fig. 3.17 Multi barrier concept of the abandonment technology for CO_2 wells under high pressure

3.2.3 Abandonment of High Pressure CO_2 Wells

The post operational phase of a CO_2 storage site begins with the abandonment of the wells under high pressure, with the reservoir filled with CO_2 to a defined final stage. For this long-term period, the well must be equipped with CO_2 resistant and high pressure stable sealing materials. The abandonment technology differs from the conventional sealing technology for gas and oil wells at low pressure. It should be determined by a multi barrier concept as presented in Fig. 3.17.

- The first barrier is a polymer gel block in the reservoir zone adjacent to the well. This gel block is formed by selected polymers as cited in Fig. 3.16, combined with the special polymer PAtBA + PEI, which is stable at temperatures until 190 °C.
- Subsequently, a gel – cement – mixture inside the well at reservoir level works as a second barrier. The given polymers were tested under CO_2 atmosphere in laboratory investigations without occurrence of structural degradation.
- The main barrier is formed by the inert plug made of CO_2 resistant and impermeable material in the well zone of the cap rock. For this plug, natural materials are assumed to be most adequate, having proved their resistance to CO_2 in nature for many millions of years.

A brief summary of the characteristics of the three materials salt, clay and sulphur is given in Fig. 3.18. Considering the stability at high temperatures, natural materials are applicable in different depths. Laboratory investigations for CO_2 stability and impermeability are currently carried out. Halite can be applied in any depths due to its resistance to high temperatures, providing the presence of a salt layer of sufficient thickness in the cap rock. Clays such as bentonite and sulphur

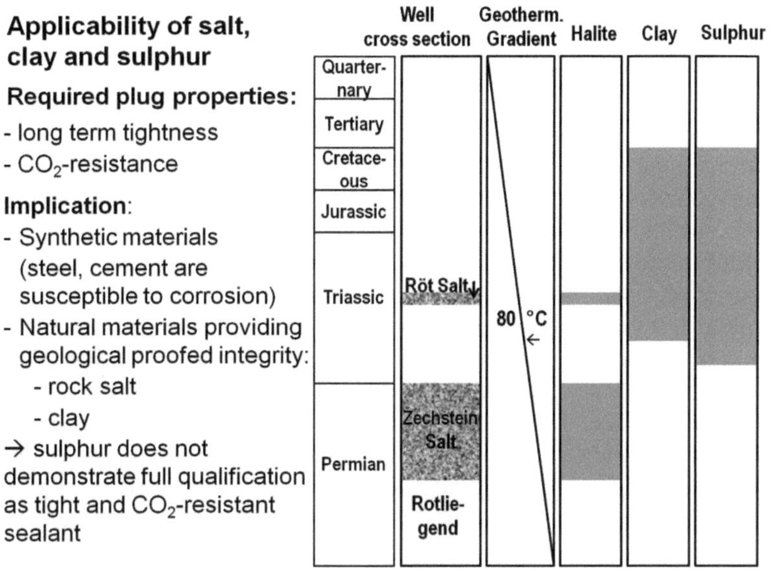

Fig. 3.18 Well abandonment with an inert, CO_2 resistant plug of natural materials

are limited to depths of 1,500–2,000 m or temperatures of 80–100 °C, accordingly. In case of higher temperatures, chemical and mineralogical stability is not given.

3.3 Development of a Long-Term Wellbore Sealing Concept

Unlike methane or nitrogen, the injected CO_2, in combination with reservoir water, forms carbonic acid, which under certain conditions may affect the integrity of wellbores in the long-term. Recommendations to improve the conventional abandonment procedure are, e.g. given in Reinicke et al. (2009a, b). The discovery of entrapped CO_2 and hydrocarbon fluid inclusions in deep rock salt formations in central and northern Germany (Siemann 2007), gives fairly good evidence for the high resistance and impermeability of rock salt under high pressure and high temperature (HTHP) conditions. Hence, rock salt is considered an ideal natural barrier, able to contain CO_2 long-term in geological storage formations. Rock salt has another important property, namely the creep deformation behaviour. Since the beginning of salt mining creep induced convergence has been observed worldwide. Convergence measurements in salt mines, for instance, dated back to the early 1960s (Bradshaw et al. 1964). It is well known that the transient and stationary creep deformations can cause a change of the salt structure without damage and dilatancy. Furthermore, excavation damaged zones (EDZ) could be sealed e.g. by backfilling the excavation rooms with crushed salt and even healed under favourable HTHP creep conditions as encountered in the Altmark gas field, where the Rotliegend sandstone reservoirs are overlain by massive plastic

Zechstein anhydrite and rock salt formations of several hundred metres thickness. The Zechstein forms a proven natural gas-tight barrier that has managed to trap natural gas for millions of years.

In the last decades, the field owner and operator, GDF SUEZ, acquired extensive technical field experiences and know-how in drilling, well completion and abandonment, with special regard to salt creep under the given geological HTHP conditions. The main measure taken to achieve enhanced borehole stability in the Zechstein rock salt section of the wellbore has been the use of double casing configurations. These experiences formed the basis and motivation to study and test the utilization of creep behaviour of rock salt formations to develop a new sealing concept.

The field test involved two major work overs. After the first work over, the volume convergence was monitored by measuring displaced wellbore fluids and solids. Results of these observations were used to verify the numerical models of borehole convergence. The second work over served for acquisition of rock salt core samples from the newly established sealing barrier in the reamed section of the wellbore. The field test was concluded with the formation integrity test, performed to prove the hydraulic integrity of the cap rock surrounding the test wellbore.

3.3.1 Multi Barrier Concept for a Long-Term Borehole Sealing System

A multi-barrier sealing concept is defined as a combination of both geotechnical and geological barriers, with specific material properties and behaviour, utilized to permanently seal boreholes on the long-term basis. Geotechnical barriers result from the emplacement and usage of sealing materials which may or may not be compatible in composition and behaviour to the virgin rock (Zechstein rock salt), while geological barriers may include the impermeable cap rock formations (e.g. shales and rock salt). The joint interaction of these barriers and their individual time dependent functions provide long-term safety for the acquired seal (Hou and Wundram 2009). As described in Fuenkajorn and Daemen (1996) the practice of sealing boreholes by using different materials (besides cement) has been investigated extensively, especially to seal boreholes of abandoned salt caverns as well as salt mines.

A combination of geotechnical and geological barriers may be successfully applied in wellbore sealing as illustrated in Fig. 3.19, which show in sequence, the stepwise procedure to develop the model for the long-term wellbore sealing concept. In phase A, a simple geological stratigraphic section of the Zechstein series in the North German Basin, particularly at the Altmark Natural Gas Field, is presented. The study selects and concentrates on the Stassfurt Zechstein (z2) rock salt formation as the most suitable candidate for the long-term barrier because of its significant homogeneity, expansiveness and vertical thickness. Therefore, it provides the ideal conditions for uniform creep and promises a high chance of recovering the former cap rock integrity. Phase A in Fig. 3.19 shows the potential period of CO_2 injection through a tubing and packer configuration. Phase B

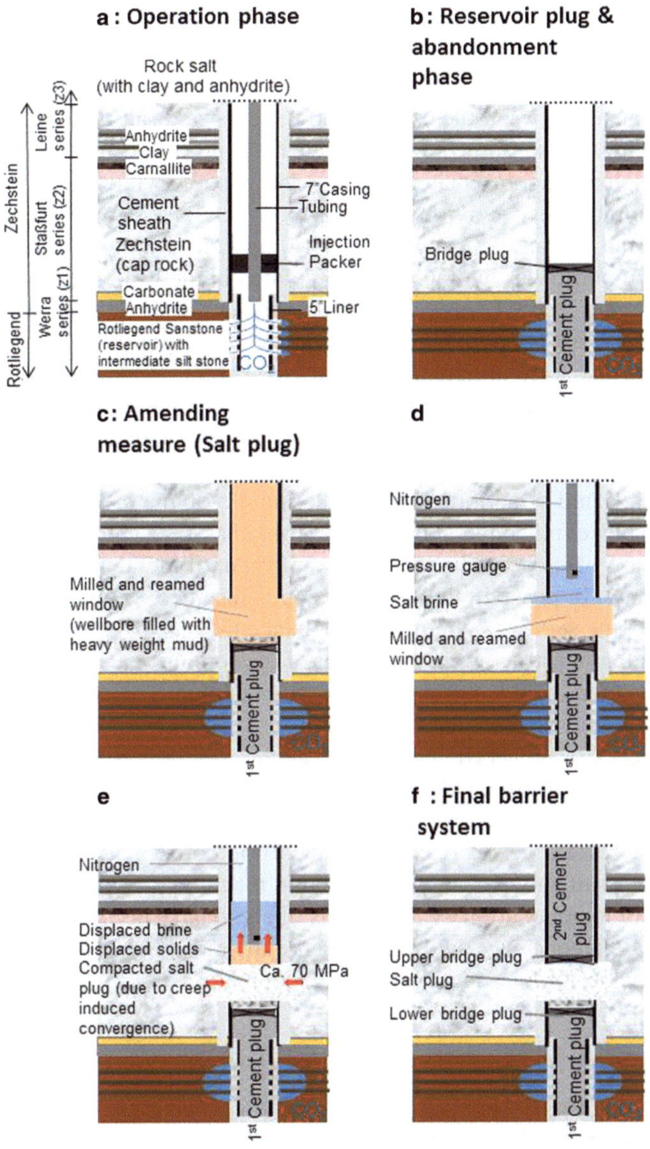

Fig. 3.19 Long-term wellbore sealing concept: (**a**) phase of CO_2 injection, (**b**) first cement bridge and bridge plug setting, (**c**) milled and reamed section of wellbore filled with heavy mud, (**d**) tubing installation; fluid replacement with brine and nitrogen to decrease the hydrostatic pressure; pressure gauge installation, (**e**) creep induced borehole convergence, initiated by surrounding Zechstein; sediment and fluid displacement, (**f**) further salt plug compaction and recrystallization in between two cement bridges (abutments); second cement bridge within upper formations

describes the installation of the first geotechnical barrier in form of a cement bridge, stretching from the Rotliegend reservoir through the thin anhydrite and carbonate layers to the Zechstein formation, and terminating with a bridge plug, a mechanical barrier (abutment) between the cement and the expected salt plug, by creep, in the reamed window. Phase C shows the geotechnical process of milling and reaming a ca. 30 m high and approximately 11 in. wide window within the 7 in. casing. In order to inhibit the creep activity of the exposed rock salt from the reamed borehole walls during the operation, heavy weight drilling fluid, with a mixed proportion of barite and bentonite, was hydraulically circulated in the wellbore. Phases D and E illustrate how the heavy mud was partially replaced with a lighter salt brine and nitrogen to decrease the hydrostatic pressure on the exposed rock salt around the reamed window so as to initiate the natural creep deformation process, responsible for the time dependent borehole convergence and the consequential formation of the compacted salt plug. Some measurements were periodically conducted in the wellbore to determine the rate of convergence, by calculating the volumes of displaced solids and brine in relation to the initial volume of the reamed window. In order to achieve full convergence, the salt plug has to be gradually compacted by steady decrease of its porosity and permeability, until the ideal properties of rock salt are restored. In phase F, the upper bridge is installed, which together with the lower bridge plug serves as abutment to enhance the tendency for recrystallization of the rock salt grain structure in the course of the coming years. Migration of fluids from the formations below and above along the salt plug is prevented by the cement bridges functioning as geotechnical barriers.

Under the existing HTHP conditions at the wellbore site, the geotechnical barriers (cement and bridge plugs) will integrally and effectively function until the geologic barrier (the compacted, crystallized salt plug) has been established in the course of several years. A comparison of the cored material with the virgin Zechstein rock salt shows that a considerable magnitude of recovery of cap rock integrity (prior to penetration by drilling) is likely to have been reached. Therefore, the tested sealing concept would ideally serve to amend/improve the conventional sealing procedure for long-term geological storage of CO_2.

3.3.2 Numerical Modelling of a Long-Term Seal

In order to describe the creep behaviour of rock salt the constitutive models LUBBY2 and Hou/Lux were applied. The Hou/Lux model (Hou 1997) is based on the LUBBY2 model (Heusermann et al. 1983) for transient and stationary creep of salt rocks and continuum damage mechanics. In comparison with the LUBBY2 model, the Hou/Lux model can additionally describe the tertiary creep, damage, dilatancy and damage healing. The details about these models are described by Hou (2002, 2003), Heusermann et al. (1983) and Hou et al. (2010b). The creep parameter sets for rock salt with the upper and lower stationary creep rate boundary are partially based on the experimental results of creep tests on salt samples from the Altmark gas field and partially on the above mentioned literatures.

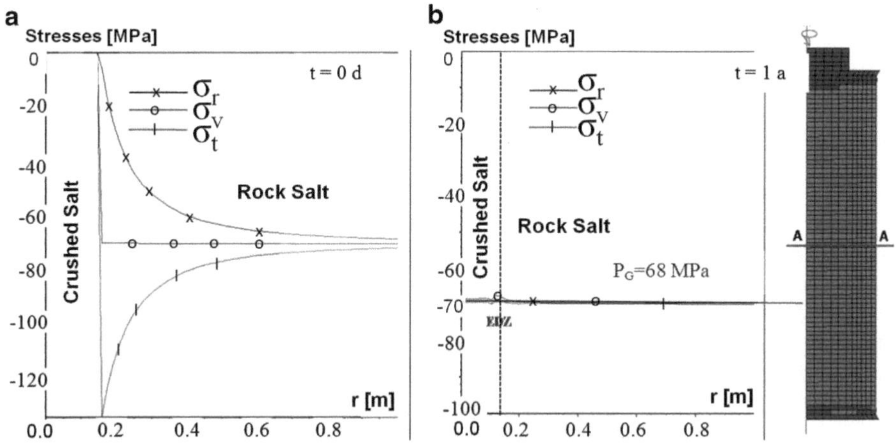

Fig. 3.20 Radial, tangential and vertical stress distributions based on the lower stationary creep boundary in crushed and rock salt mass: (**a**) immediately after the insertion of crushed salt, and (**b**) after 1 year

3.3.2.1 Numerical Simulation with a Backfill

Prior to the execution of the field test numerical simulations, as pre-calculations for creep induced compaction in crushed salt, were performed under application of the CWIPP constitutive model in Sjaardema and Krieg (1987) as a cost-effective method to predict the time period of backfill density alignment with the surrounding rock salt. After completion of the reaming job and crushed salt backfill in place, the primary stress state of rock salt is translated into the secondary stress state, whereby the three stress components (radial, tangential and vertical) are initially set to be zero in the backfill, while the highest load is at the borehole walls (Fig. 3.20 left). It is observed that creep induced borehole convergence leads to the time dependent gradual compaction resulting in a stress build-up in the backfill until the primary stress state is recovered after approximately 1 year (Fig. 3.20 right).

3.3.2.2 Numerical Simulation Without a Backfill

In the second run of simulations evolution of borehole volume convergence was investigated by using the Hou/Lux constitutive model for rock salt, including the creep induced dilatancy, damage healing and permeability in the reamed section of the wellbore without a backfill. The results are shown in Fig. 3.21.

In Fig. 3.21 (left hand) a gradual decay or increase of stresses in the borehole (until the vertical dark dashed line) after 5 years of creep, and a significant increase of stresses in the EDZ is observed. The simulation predicts a longer time duration ($\gg$5 years) until the primary stress of ca. 70 MPa will be reached. The time

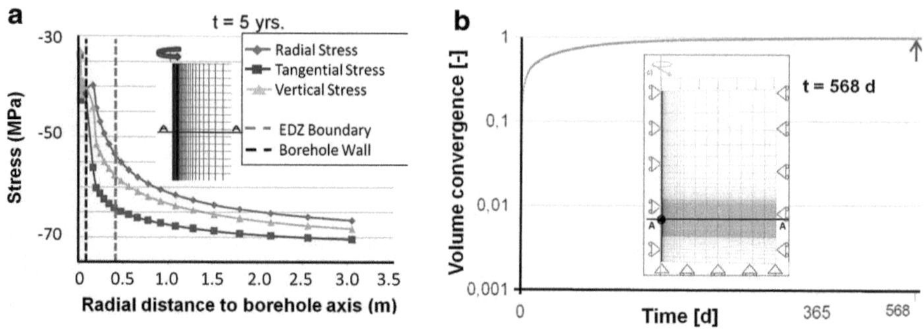

Fig. 3.21 (**a**) radial, tangential and vertical stresses in the mid-horizontal section of the reamed window after 5 years of convergence/creep period. (**b**) volume convergence using the Hou/Lux model based on the lower stationary creep boundary

dependant volume convergence is shown in Fig. 3.21 (right hand), and reveals a situation of almost complete convergence after ca. 1.5 years.

As a result, the time duration for sealing the borehole without backfill is longer than in presence of a backfill. However, it must be mentioned that the simulations do not consider the possible spalling due to strength exceeding at the borehole wall and thus the effects of self-backfilling are not considered. This self-backfilling and the adjacent EDZ would be compacted and even healed under the existing HTHP conditions. Nevertheless, in any case the time duration should be longer without backfilling by crushed salt in the reamed section.

3.3.3 Field Test

The selected well for the field test is located in an area where the Zechstein rock salt formation is relatively thick and homogenous. Field work in the study wellbore was conducted in three phases, each with its own specific objectives:

- Phase I (first work over): This phase marked the start of the field test in March 2010, and lasted for 1 month and comprised creation of a window exposing the Zechstein rock salt formation in the wellbore by cutting and reaming the casing in the corresponding section of the wellbore (Fig. 3.19, phases B and C).
- Phase II (observation of convergence): This phase followed immediately after the exposure of the rock salt in the wellbore in April 2010 and lasted more than 1 year, until June 2011 (ca. 450 days). Its main objective was to observe the time dependant development of volume convergence and wellbore integrity in the reamed window by regularly taking brine level measurements and samples.
- Phase III (second work over): The last phase lasted about 2 weeks after the end of the second phase and was mainly concerned with coring and testing the integrity of the formation in and around the window section of the wellbore. This phase ended with the application of conventional abandonment procedure on the test wellbore above the reamed window.

3.3.3.1 The First Work over: Phase I

The primary objective of the field test was to investigate the technical feasibility to induce Zechstein salt creep and therefore to seal a formerly reamed section of the wellbore, where casing and cement had been removed. Considering the simulation results, it was expected that the initial cap rock integrity would be restored in a relatively short period of time.

The initial activities of the work over comprised the following sequence of procedures: (1) setting a cement plug in the Rotliegend reservoir in order to secure operation safety, (2) installation of a bridge plug in the space between the Rotliegend and Zechstein formations, according to the conventional standard abandonment procedure, and (3) milling and underreaming a 30 m long vertical section of casing within the thick Zechstein salt.

3.3.3.2 Observation of the Convergence: Phase II

In the observation phase, the rising brine level was regularly measured with the aim to monitor and estimate the progress of volume convergence as a result of rock salt creeping into the reamed window.

Daily measurements and observations of the brine and solid particle level, including temperature and pressure at the pressure gauge level were taken, during the first 2 weeks and periodically every 8–10 weeks afterwards until the beginning of phase III (Fig. 3.22). Furthermore, this phase included a hydraulic pressure test, laboratory tests on fluid and sediment samples, back-analysis of the solid particle level as well as evaluation of the volume convergences (Fig. 3.23).

From the periodical measurements of brine levels, the volume displaced by the brine was related to the initial volume of the reamed window. This analysis is based on the displaced sediments observed above the reamed section.

In consideration of 53 % sediment porosity (bentonite) from laboratory analysis and a convergence volume of ca. 2.3 m^3 instead of 2.4 m^3 due to remaining solids in the reamed section, a volume convergence of ca. 50 % was calculated within ca. 16 days after the beginning of measurements. It can be seen that the measured volume convergence (ca. 50–94 %) during the observation period ($t \approx 400$ days) evolves linearly with time while the slope decreases slightly after ca. 300 days (Fig. 3.23).

3.3.3.3 The Second Work over: Phase III

The aim of the second work over phase was to obtain core material of the new re-crystallized rock salt, formed as a result of convergence, within the originally reamed window. The procedure involved replacing the compressed nitrogen in the wellbore with drilling mud and the subsequent removal of solid particles from the casing section above the reamed window. During the flushing operation, a collapse of the casing was detected directly above the reamed window. The

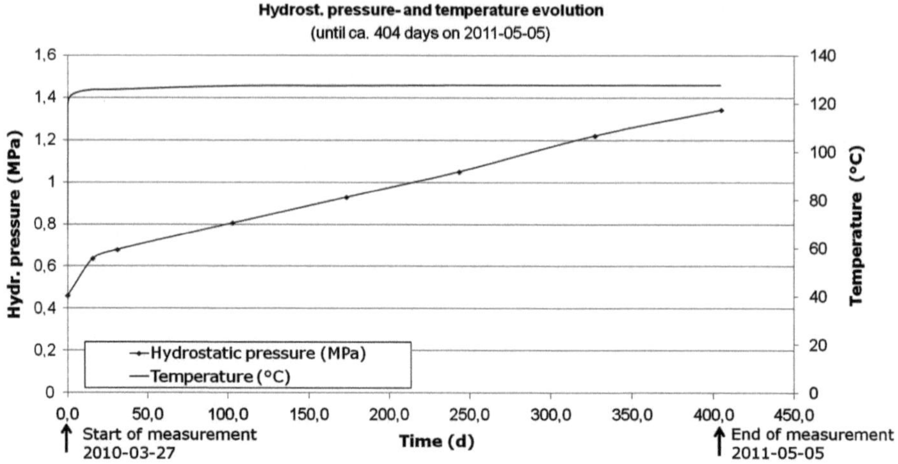

Fig. 3.22 Measured hydraulic pressure and brine temperature as function of time

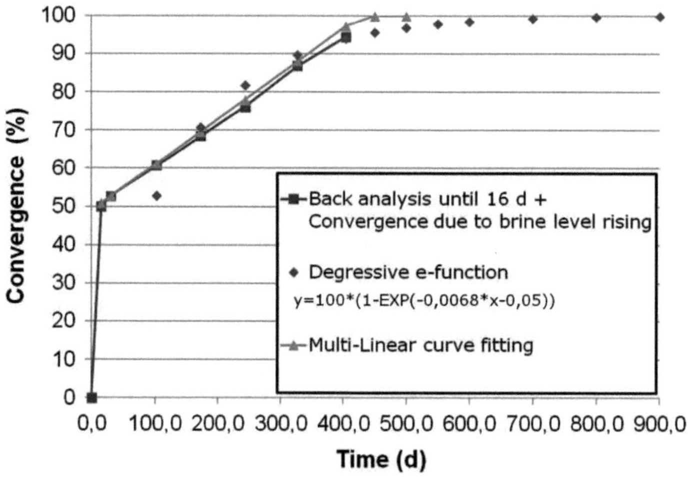

Fig. 3.23 Evolution of the in-situ volume convergence curve with explicit values and the pre-calculated fitting curve (through exponential and multi-linear functions) as a function of time. The graphs can be used to predict total convergence

collapsed casing was successfully eliminated by a taper milling operation. Core material of good quality for further lab analysis was obtained.

With the first visual inspection of the cored material it was observed that a re-crystallized salt structure and highly consolidated (compacted) bentonite/barite mud sediments were already accumulated in the convergence zone. The salt/sediment interface clearly reveals that the coring was accomplished within the radius of the formerly milled section and hence, in the newly formed rock salt material.

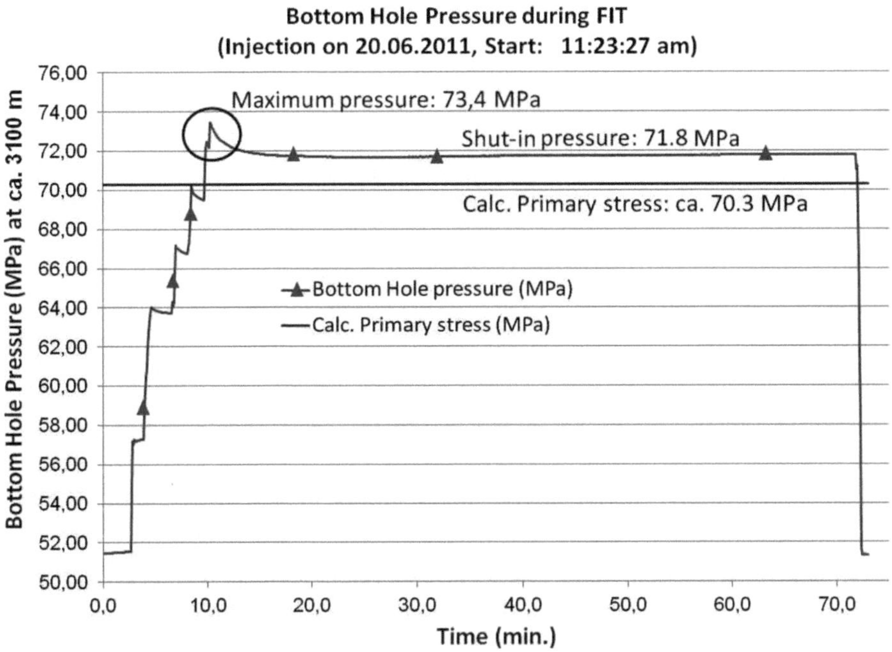

Fig. 3.24 Evolution of bottom hole pressure (BHP) as a function of time during a formation integrity test (FIT). Notice the stepwise increase of pressure until the maximum value of 73.4 MPa and the slight decrease to 71.8 MPa, the shut-in pressure

It can be concluded that after having the core material at hand and under consideration of the casing collapse proven by the work over, a high convergence rate and hence, a fairly well progressed closure of the milled window did occur.

In order to approve the hydraulic tightness of the Zechstein salt formation in the direct vicinity of the test wellbore, a pressure test with a maximum pressure of ca. 73 MPa was performed after the coring operation. Figure 3.24 shows the evolution of the bottom hole pressure (BHP) with time during the formation integrity test (FIT). The measured constant shut-in pressure of 71.8 MPa during the whole shut-in phase of ca. one hour reveals the tightness/integrity of the remaining salt plug in the reamed window and the surrounding rock salt formation. Additionally, the final shut-in pressure was nearly of the same magnitude as the corresponding primary stress, indicating an effective sealing capability of the rock salt. Further details about leak-off tests in general are given in Fjaer et al. (2008).

3.3.3.4 Comparison of In-Situ Measurements and Numerical Results

Numerical simulation results conducted prior to the field test (Fig. 3.21 right hand) show that complete convergence in the reamed window without a backfill would be achieved already after ca. 1.5 years from the time of reaming. During the

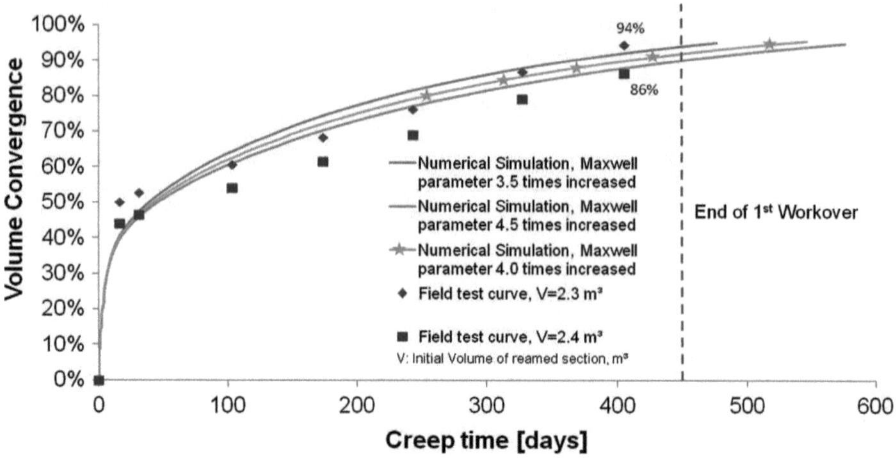

Fig. 3.25 Evolution of volume convergence as a function of time calculated based on a numerical sensitivity study and compared with the derived in-situ volume convergence

observation and measurement phase, it was observed that the derived in-situ convergence curve progresses at a slower rate than the one predicted numerically. It was therefore decided to perform several different numerical runs to find an optimal solution for the in-situ situation.

Only a reduced stationary creep rate led to a successful validation. For this reason the stationary creep parameter (Maxwell parameter, Heusermann et al. 1983) was increased to reach curve alignment. Considering all simulations done, only those in which the stationary creep parameter was increased by a factor of 3.5–4.5 led to an almost precise fit between the simulation and the field test curve (Fig. 3.25). In that way a volume convergence of ca. 90–95 % can be numerically predicted to the end of the field test (t = ca. 450 days).

3.3.3.5 Field Test Interpretation

Measurements during the two work overs and the intermediate observation phase (some 14 months in total) leave no doubt about the technical feasibility of the concept tested. The observed casing collapse, proven by milling operations approves the strong convergence of the salt rock in the vicinity of the milled window. This result is so far in line with comprehensive field experience made during former drilling and production operations in the Altmark field. The successfully commissioned formation integrity test gave additional confirmation that the system around the newly formed salt plug is hydraulically tight.

The results obtained provide valuable information about the possibility to technically take advantage of the natural mechanical properties of the Zechstein salt (the tendency to creep) with regard to amending and improving conventional abandonment procedures for wellbores and hence, increase mining safety. Numerical simulations performed show that long-term borehole sealing with or without backfill can be attained by salt creep, including the subsequent recovery of the initial integrity.

Based on the results of this study it can be concluded that the developed concept can be adapted to other wellbores under similar geologic conditions. However, the pressure conditions should be precisely determined in order to choose the right backfilling material and to allot enough time for convergence.

References

Abdel-Haq A (2006) Erfahrung bei der Durchführung des Dichtheitstestes an den letzten zementierten Rohrtouren der Kavernenfelder EPE und Bad Lauchstädt. 57. Berg- und Hüttenmännischer Tag, 22–23 Juni 2006

Bachu S, Bennion DB (2009) Experimental assessment of brine and/or CO_2 leakage through well cements at reservoir conditions. Int J Greenhouse Gas Control 3:494–501

Barlet-Gouédard V, Rimmelé G, Porcherie O, Quisel N, Desroches J (2009) A solution against well cement degradation under CO_2 geological storage environment. Int J Greenhouse Gas Control 3:206–216

Berest P, Brouard B, Durup G (2002) Tightness tests in salt-cavern wells. Technical paper. Solution Mining Research Institute, Spring 2002 Meeting, 28 April–1 May 2002, Banff. http://brouard-consulting.com/articles/tightness-banff.pdf

Bradshaw RL, Boegly WJ, Empson FM (1964) Correlation of convergence measurements in salt mines with laboratory creep-test data. In: The 6th U.S. symposium on rock mechanics (USRMS), 28–30 Oct 1964, The University of Missouri Rolla, Rolla, pp 501–514

Bruckdorfer RA (1986) Carbon dioxide corrosion in oilwell cements. Society of Petroleum Engineers SPE 15176

Carey JW, Wigand M, Chipera SJ, Wolde-Gabriel G, Pawar R, Lichtner PC, Wehner SC, Raines MA, Guthrie GD Jr (2007) Analysis and performance of oil well cement with 30 years of CO_2 exposure from the SACROC Unit, West Texas, USA. Intl J Greenhouse Gas Control 1(1):75–85

Condor J, Asghari K (2009) An alternative theoretical methodology for monitoring the risks of CO_2 leakage from wellbores. Energy Procedia 1:2599–2605

Crow W, Carey JW, Gasda S, Williams DB, Celia M (2010) Wellbore integrity analysis of a natural gas producer. Int J Greenhouse Gas Control 4:186–197

Fjaer E, Holt RM, Horsrud P, Raaen AM, Risnes R (2008) Petroleum related rock mechanics. Dev Petr Sci 53:299–304

Fuenkajorn K, Daemen JJK (1996) Sealing of boreholes und underground excavations in rock. Chapman & Hall, London, pp 184–211

Hendricks (2009) Experiences in the Salt Creek Field CO_2 flood. In: 5th annual wellbore integrity network, Calgary, 13–14 May 2009

Heusermann S, Lux KH, Rokahr R (1983) Entwicklung mathematisch-mechanischer Modelle zur Beschreibung des Stoffverhaltens von Salzgestein in Abhängigkeit von der Zeit und der Temperatur auf der Grundlage von Laborversuchen mit begleitenden kontinuumsmechanischen Berechnungen nach der Methode der finiten Elemente. BMBF-Forschungsbericht T 83–218. Lehrgebiet Unterirdisches Bauen der Universität Hannover

Hou Z (1997) Untersuchungen zum Nachweis der Standsicherheit für Untertagedeponien im Salzgebirge. Dissertation an der TU Clausthal, Papierflieger Verlag in Clausthal-Zellerfeld

Hou Z (2002) Geomechanische Planungskonzepte für untertägige Tragwerke mit besonderer Berücksichtigung von Gefügeschädigung, Verheilung und hydromechanischer Kopplung. Habilitationsschrift an der TU Clausthal, Papierflieger Verlag in Clausthal-Zellerfeld

Hou Z (2003) Mechanical and hydraulic behaviour of salt in the excavation disturbed zone around underground facilities. Int J Rock Mech Min Sci 40(5):725–738

Hou Z, Wundram L (2009) Langzeitverschluss der CO_2-Entsorgung in einer Erdgaslagerstätte zur Erhöhung der Ausbeutungsrate: Vergleich mit der Endlagerung radioaktiver Abfälle in einem alten Bergwerk. Energie und Rohstoffe, Goslar, 9–12 Sept 2009, pp 317–334

Hou Z, Wang Q, Xie L, Liu J, Zhang R (2010) The Hou's short- and long-term strength model, the Hou/Lux constitutive model for rock salt as well as their applications in storage cavern design. In: Hou Z, Xie H, Yoon JS (eds) Underground storage of CO_2 and energy. CRC Press/Balkema, Leiden, pp 95–103

ITE (2007) Feasibility study on the potential of CO_2 storage for enhancing recovery in mature German gas reservoirs (CSEGR) – work package 2: CO_2 injection/production and well integrity. Beitrag zum Abschlussbericht des Gemeinschaftsvorhabens unter dem BMBF Sonderprogramm GEOTECHNOLOGIEN

Jacobsen S, Marchand J, Gerard B (1998) Concrete cracks I: durability and self healing. In: Gjorv OE, Sakai K, Banthia NE (eds) Proceedings of the second international conference on concrete under severe conditions. E & FN Spon, Tromso, pp 217–231

Krilov Z, Loncaric B, Miksa Z (2000) Investigation of a long-term cement deterioration under a high temperature sour gas downhole environment, Society of Petroleum Engineers. The 2000 SPE international symposium on formation damage control, Lafayette, 23–24 Feb 2000, Paper no. SPE 58771, pp 547–555

Mazurek M, Pearson FJ, Volckaert G, Bock H (2003) Features, events and processes evaluation catalogue for argillaceous media. Radioactive Waste Management, ISBN 92-64-02148-5. © OECD 2003

Nygaard R (2010) Well design and well integrity, Wabamun area CO_2 sequestration project (WASP), Energy and Environmental Systems Group, Institute Sustainable Energy, Environment and Economy (ISEEE), University of Calgary, p 39

Quintessa (2011) CO_2 FEP data base. http://www.quintessa-online.com/co2/

Reinicke KM, Fichter C (2009) CO_2 Lagerung im Geogrund: Bohrungsintegrität akzeptabel für CSEGR (carbon sequestration with enhanced gas recovery) Part 1 and 2. Erdöl-Erdgas-Kohle, Januar und Februar 2009

Reinicke KM, Fichter C (2010a) Measurement strategies to evaluate the integrity of deep wells for CO_2 applications. Proceedings of the Sino-German conference on underground storage of CO_2 and energy, Beijing, 6–13 July 2010

Reinicke KM, Fichter C (2010b) Entwicklung von Prüfstrategien zur Bewertung der technischen Integrität von Bohrungen. DGMK/ÖGEW-Frühjahrstagung 2010, Fachbereich Aufsuchung und Gewinnung, Celle, 12–13 Apr 2010

Reinicke KM, Weichmann MJ (2011) Risikobewertung verfüllter Bohrungen unter dem Einfluss von CO_2. Freiberger Forschungsforum 2011, 8. Fachkolloquium GEO Förderung und Speicherung von Energie im Untergrund, 15–17 Juni 2011

Reinicke KM, Franz O, Fichter C (2009a) Bohrungsintegrität akzeptabel für CSEGR CO_2-Lagerung im Geountergrund (1). Erdöl Erdgas Kohle 125(1):27–31

Reinicke KM, Franz O, Fichter C (2009b) Bohrungsintegrität akzeptabel für CSEGR CO_2-Lagerung im Geountergrund (2). Erdöl Erdgas Kohle 125(2):66–67

Reinicke KM, Bai M, Fichter C, Weichmann MJ, Weinlich FH (2011a) Leckagerisiko von Bohrungen unter Einfluss von CO_2: Von den FEPs bis zu den Szenarien. DGMK/ÖGEW-Frühjahrstagung 2011, Celle, 11–12 Apr 2011

Reinicke KM, Teodoriu C, Fichter C, Weichmann MJ, Weinlich FH, Krebs R, Zhang Z-L (2011b) Untersuchung von Selbstheilungsmechanismen im Verbundsystem Rohr-Zement-Gebirge von CO_2-Bohrungen. DGMK/ÖGEW-Frühjahrstagung 2011, Celle, 11–12 Apr 2011

Schlangen E, Rooij M de (2005) Self-healing phenomena in cement-based materials. International union of laboratories & experts in construction materials, systems and structures, technical committee 221-SHC. http://www.rilem.net/gene/main.php?base=8750&gp_id= 228

Schmitt G, Zhang Y, Gao K (2011a) Water effect on steel corrosion under supercritical CO_2 conditions, NACE CORROSION'2011, Houston, 13–17 Mar 2011, Paper 11378

Schmitt G, Zhang Y, Gao K (2011b) Water effect on steel corrosion under supercritical CO_2 conditions. Mater Perform 50(6):62

Schmitt G, Zhang Y, Gao K (2011c) Inhibition of steel corrosion under aqueous supercritical CO_2 conditions, NACE CORROSION'2011, Houston, 13–17 Mar 2011, Paper 11379
Schmitt G, Zhang Y, Gao K (2011d) Inhibition of steel corrosion under aqueous supercritical CO_2 conditions. Mater Perform 50(9):54
Schmitt G, Zhang Y, Gao K (2011e) Corrosion inhibition of steels in aqueous media in contact with supercritical carbon dioxide, EUROCORR'2011, 4–8 Sept. 2011, Stockholm, Paper 4717
Schmitt G, Zhang Y, Gao K (2011f) Effect of oxygen on steel corrosion under supercritical CO_2 conditions, EUROCORR'2011, 4–8 Sept. 2011, Stockholm, Paper 1103
Schmitt G, Zhang Y, Gao K (2011g) Corrosion inhibition of CRAs in systems with supercritical carbon dioxide, 18th international corrosion congress, 20–24 Nov 2011, Perth, Paper 82
Schmitt G (2011h) Stähle in überkritischen CO_2 Systemen – Unerwarteter Sauerstoff-Effekt, GfKORR-Workshop: CCS-Technologie – Werkstofffragen prozesstechnisch lösen, 14 Dec 2011, DECHEMA, Frankfurt/Main
Schmitt G, Zhang Y, Gao K, Hausler RH (2012) Modeling steel corrosion under supercritical CO_2 conditions. Mater Corrosion 63(2). doi:10.1002/maco.201106382, in press
Siemann MG (2007) Herkunft und Migration mineralgebundener Gase der Zechstein 2 Schichten in Zielitz. Kali Steinsalz 3:26–41
Sjaardema GD, Krieg RD (1987) A constitutive model for the consolidation of crushed salt and its use in analyses of backfilled shaft and drift configurations. SAND87-1977. Sandia National Laboratories, Albuquerque
SMRI (1996) SMRI reference for external well mechanical integrity testing/performance, data evaluation and assessment; SMRI Spring meeting, Crotogino F., Short Class, Houston
Teodoriu C, Reinicke KM, Fichter C, Wehling P (2010) Investigations on casing-cement interaction with application to gas and CO_2 storage wells. SPE 131336, 2010 EAGE/SPE EUROPEC, 14–17 June 2010, Barcelona
Tran Viet T (2010) Bewertung Technische Integrität: Bewertung bestehender Verfahren, In: CLEAN Zwischenbericht Dezember 2009 und CLEAN Jahrestreffen Januar 2010
Williams S, Carlsen T, Constable K, Guldahl G (2009) Identification and qualification of shale annular barriers using wireline logs during plug and abandonment operations. SPE/IADC 119321, SPE Well Abandonment, Aberdeen, 30 June 2009. http://www.spe-uk.org/Downloads/Well-Abandonment-09/1530_Shale_Annular_Barriers_E_Gudding_StatoilHydro.pdf
Zhang C-L, Müller J, Ziesche U (2011) Untersuchungen zur Integrität von Rohr-Zement-Salz. Teilprojekt des Verbundvorhabens EGR Altmark CO_2SAFE. Unveröffentlichter GRS-Laborbericht im Auftrage des ITE (TUC)

Chapter 4
Evaluation of Geo-processes

Andrea Förster, Daniel Albrecht, Sebastian Bauer, Gunther Baumann, Christof Beyer, Norbert Böttcher, Roland Braun, Knut Behrends, Ronald Conze, Marco De Lucia, Leonhard Ganzer, Reinhard Gaupp, Uwe-Jens Görke, Yang Gou, Jan Henninges, Zengmeng Hou, Bernd Kohlhepp, Olaf Kolditz, Michael Kühn, Christof Lempp, Rudolf Liedl, Robert Meyer, Ben Norden, Thomas Nowak, Peter Pilz, Dieter Pudlo, Matthias Rateizak, Viktor Reitenbach, Khaled M. Shams, Haibing Shao, Ashok K. Singh, Susanne Stadler, Lutz Stecken, Joshua Taron, Wenqing Wang, Norihiro Watanabe, Joachim Wienand, Jeoungseok Yoon and Martin Zimmer

A. Förster (✉) • G. Baumann • K. Behrends • R. Conze • M. De Lucia • J. Henninges • M. Kühn • B. Norden • P. Pilz • M. Zimmer
Helmholtz Centre Potsdam, GFZ – German Research Centre for Geosciences, Telegrafenberg, 14473 Potsdam, Germany
e-mail: for@gfz-potsdam.de

D. Albrecht • L. Ganzer • V. Reitenbach • J. Wienand
Institute of Petroleum Engineering, Clausthal University of Technology, 38678 Clausthal-Zellerfeld, Germany

S. Bauer • C. Beyer
CAU Christian-Albrechts-Universität of Kiel, Ludewig-Meyn-Str. 10, 24118 Kiel, Germany

N. Böttcher • R. Liedl
Technical University Dresden, 01062 Dresden, Germany

R. Braun
Dr. Roland Braun – Consultancy in Rock Mechanics, 14548 Caputh, Germany

R. Gaupp • B. Kohlhepp • D. Pudlo
Institute for Geosciences, Friedrich Schiller University, 07749 Jena, Germany

U.-J. Görke • H. Shao • A.K. Singh • J. Taron • W. Wang • N. Watanabe
Helmholtz Centre for Environmental Research – UFZ, Permoserstr. 15, 04318 Leipzig, Germany

Y. Gou • J. Yoon
Lower Saxony Energy Research Centre (EFZN) 38640 Goslar, Germany

Z. Hou
Institute of Petroleum Engineering, Clausthal University of Technology, 38678 Clausthal-Zellerfeld, Germany

Lower Saxony Energy Research Centre (EFZN), 38640 Goslar, Germany

M. Kühn and U. Münch (eds.), *CLEAN: CO_2 Large-Scale Enhanced Gas Recovery in the Altmark Natural Gas Field*, Advanced Technologies in Earth Sciences,
DOI 10.1007/978-3-642-31677-7_4, © Springer-Verlag Berlin Heidelberg 2013

Abstract A holistic understanding of the physicochemical processes induced by CO_2 injection and storage in a reservoir is based on a geoscientific characterisation of the overall geological system consisting of reservoir rocks and cap rocks. It requires in a first step a comprehensive baseline characterisation (sedimentological, mineralogical, geochemical, mechanical, etc.) of pertinent parameters and conditions. To properly handle the large amount of different geoscientific information a Data Management System (DMS) was developed, which proved indispensable to conduct such a multi-disciplinary project. The DMS provides a tool for scientific process management, data analysis, integration and visualisation, data transfer and scheduling through specialised database systems and retrieval techniques, storage technology, and efficient data access.

Sedimentological (facies), mineralogical and petrophysical data classify the Altmark Rotliegend sandstones of high quality with best porosity and permeability in altered/bleached aeolian sandstones. Those properties are strongly controlled by sediment deposition (facies-type) and by early and late diagenesis, namely by early pore-filling cementation and late fluid-rock interaction. The fluids involved probably originated from Carboniferous formations and ascended along (re-) activated faults through the Rotliegend volcanic rocks during Triassic-Jurassic times. Major features controlling the extent of fluid-rock reactions are grain size and grain sorting, respectively pore size, pore shape and pore connectivity. Especially grain coating chlorite and pore-filling carbonate and anhydrite will most probably affect reactivity during CO_2 injection. Laboratory batch experiments with core samples reveal that CO_2 saturated brines cause dissolution of (pore-filling) minerals and alteration of fluid and flow properties of the reservoir rocks. This also was emphasised by a parameter study based on numerical modelling, which indicate a tendency of increasing anhydrite dissolution and calcite precipitation with increasing CO_2 partial pressure. Increasing temperature and salinity counteracts this effect. Changes of rock properties observed comprise an increase in porosity, permeability, water binding capacity, rock wettability and a slight decrease in residual gas saturation.

The recent stress field was determined by direct measurement of the effective primary principal stresses. The total primary stresses were calculated from the effective primary stresses, pore-pressure effectiveness and pore pressure. The orientations of the primary horizontal stresses coincide with the directions of the fault zones in the

O. Kolditz
Technical University Dresden, 01062 Dresden, Germany

Helmholtz Centre for Environmental Research – UFZ, Permoserstr. 15, 04318 Leipzig, Germany

C. Lempp • K.M. Shams
Institute for Geosciences, Martin-Luther-University Halle-Wittenberg, 06099 Halle/Saale, Germany

R. Meyer • L. Stecken
GDF SUEZ E&P Deutschland GmbH, Waldstrasse 39, 49808 Lingen

T. Nowak • M. Rateizak • S. Stadler
Federal Institute for Geosciences and Natural Resources (BGR), 30655 Hannover, Germany

Salzwedel area. Stress ratios for the anhydrite and claystone strata, calculated from the obtained effective primary stresses, are expected to be modified with a pilot injection into the reservoir.

Methods and numerical tools were developed, which are dedicated to the numerical characterisation of the nearly depleted gas reservoir as well as to the simulation of the processes during CO_2 injection and migration storage. The only practical option for predicting the long-term behaviour of CO_2 in reservoirs is numerical analysis, supported by the understanding gained from the relatively short-term laboratory and field-scale experiments. Corresponding to the real site conditions, compositional gas flow has to be considered including mutual interactions with thermal, mechanical and geochemical (reactive) processes taking into account high pressures and temperatures. Process- as well as site-related benchmarks were jointly developed and implemented. It is observed that the high accuracy of complex equations of state (EoS) do not justify the higher computing costs compared to simple, cubic equations of state. The Peng-Robinson equation of state is the most suitable EoS with regard to pure CO_2. At no stage of the process there is any evidence of plastic deformation. Both reservoir and cap rock behave elastically. In general, CO_2-EGR operation might cause only local changes in the pore pressure of the reservoir, while CO_2 storage will increase the reservoir pressure dependent on the injection rate if no gas is produced at the same time. Simulations revealed that an initial tensile stress regime is the safest precondition while the compressive stress state is problematic with regard to the reservoir integrity. The calculated, virtual tracer test, applying krypton, indicates that concentrations at the monitoring wells would be too low to be detected. Hence a test under the considered constraints is not feasible.

4.1 Geoscientific Integrated Data Management System

Already in the early stage of the CLEAN project it became obvious that the thematic network “Geo-Processes” would be confronted with a wealth of historical and new data that required special attention in their assessment and storage, but also in sharing them in the generation of new models to be produced in the course of the project. Considering these needs, a Data Management System (DMS) was developed as an infrastructure. During the project, the DMS has proven evidently that a tailored data management is indispensable in large temporary research projects, in which industry collaborates with academia in a multidisciplinary context.

The DMS is a web-based information and data management system in which, on the one hand, historical geoscientific exploration data archived by GDF SUEZ E&P Deutschland GmbH (GDF SUEZ) are scrutinized and screened for implementation into the CLEAN research. On the other hand, the system also compiles, screens, integrates, and stores the new data and models generated during the project. Most important, the DMS also serves as a collaboration platform, which required, due to the involvement of different sub-disciplines of geoscience and the specialised

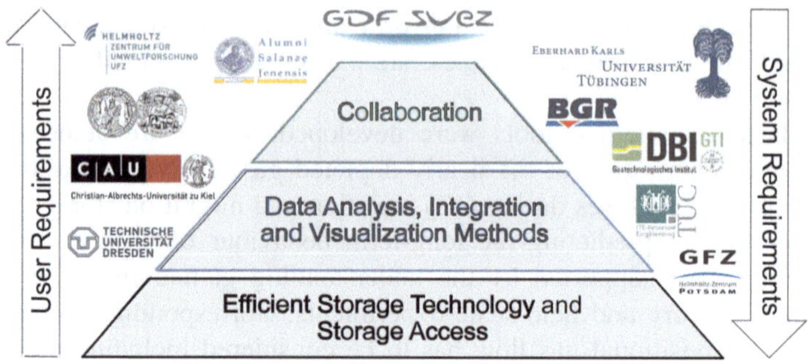

Fig. 4.1 Layered representation of DMS features and collaborating partners

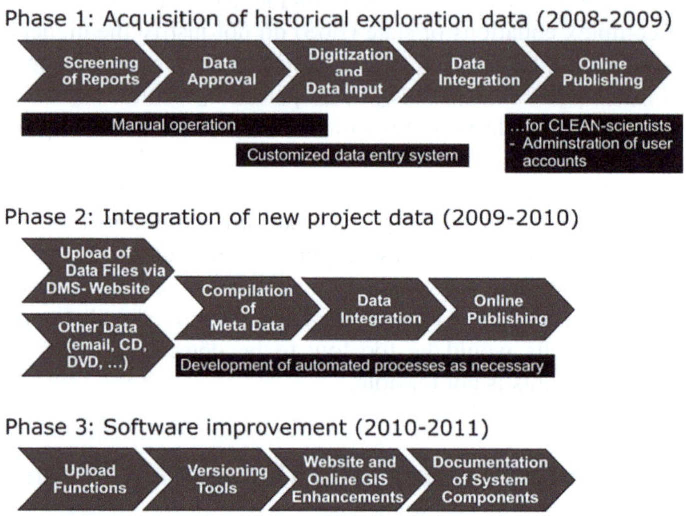

Fig. 4.2 DMS development phases

knowledge associated with unique data handling, work structures beyond the capabilities of commercial-off-the-shelf software products (Fig. 4.1).

Work on the DMS was designed for a 3-years period, starting in 2008, and was structured into three major phases (Fig. 4.2). The final phase terminated with a software documentation of the systems components. The achievement of the internal goals of the project was guided by several pre-defined milestones:

- Analysis of user requirements.
- Building the DMS data mart (data scrutinizing and digitizing).
- Integration of new data (the CLEAN data).
- Conditioning of legacy exploration data and the laboratory measurement of thermal and hydraulic properties for input into a 3D regional geological model covering the reservoir and its overburden. Generation of a 3D regional thermal model for the larger Altensalzwedel area, based on the 3D geological model.

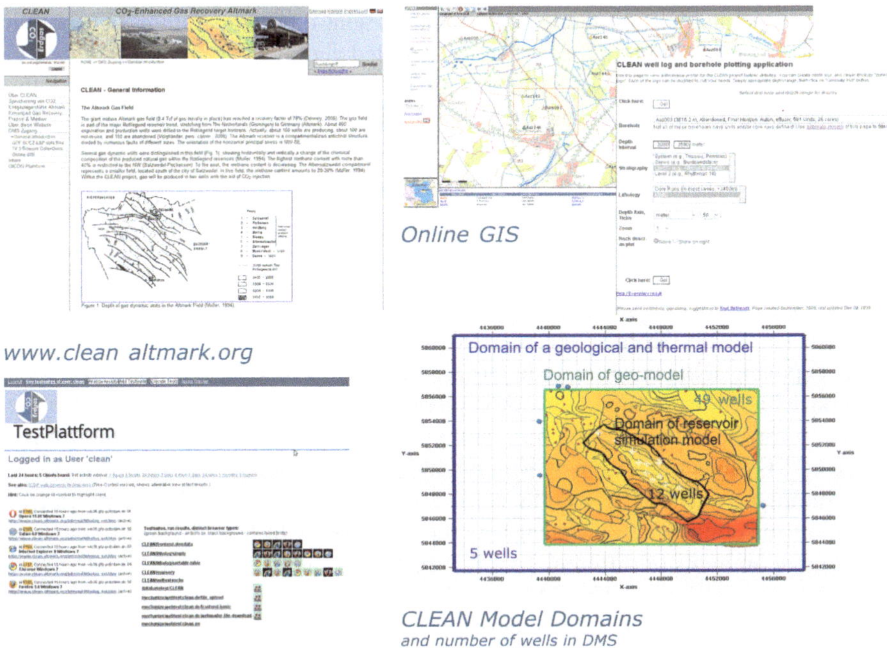

Fig. 4.3 Online data availability test suites

- Design of concepts on how to store and distribute different versions of simulation models.
- Design of concept schemes for different types of geo-models and for an archive.
- Specification and adjustment of export filters, data models and online tools.
- Documentation of system components for administrators.

The archive "exploration data" contains a huge amount of legacy data from the Altensalzwedel sub-field, but also data from the larger Altensalzwedel area. These data consist of borehole reports, structural maps, and other regional maps. In addition, lithological descriptions, laboratory porosity/permeability data, and logging data from 70 boreholes as well as traces and data from 87 seismic lines are incorporated in the DMS. They built the basis for models of different regional extent pertaining either to the reservoir or both the reservoir and its overburden. The DMS also includes an online GIS for fast geographic orientation and borehole navigation (Fig. 4.3). The new data obtained during the CLEAN project and stored in the DMS comprise mineralogical, chemical, thermal, and hydraulic data from laboratory analyses and from laboratory experiments as well as interpreted compilations of input data for modelling and process simulation, e.g. reservoir models at reservoir (far-field) and near-borehole scale as well as models at borehole scale.

A special facet of work performed during the lifetime of the DMS project was the generation of a thermal 3D model for the larger Altensalzwedel area (of the size of 350 km^2). This thermal model shows the present-day temperature distribution from a depth of 10 km up to the surface. A detailed geological structure, mapped in 3D by GDF SUEZ using a wealth of seismic reflection data, forms the basis for the thermal model. To assign thermal properties to the 14 thermal units distinguished in the model, lithological descriptions of 67 wells were analysed and a mean rock composition was delineated. The rock thermal properties assigned to these 14 units are based on data from petrophysical well-logs and laboratory measurements. Due to the very detail of legacy data and the comprehensive set of petrophysical data, the 3D geothermal model of the larger Altensalzwedel area (Norden et al. 2012) can be considered as a novel approach for showing the geothermal situation in the North-east German Basin. Even if the thermal model is regarded as a reconnaissance model exposing some uncertainty in terms of absolute temperatures, it forms a good basis on which local temperature prognoses can be made. Those predictions are necessary for the definition of temperature paths for planned boreholes and for the design of CO_2 injection regimes, which both imply important engineering aspects.

In summary, the CLEAN DMS proved indispensable to conduct a project of such multi-disciplinary nature as it was the situation with the CLEAN project, in which large amounts of different geoscientific information needed to be integrated into different approaches. The DMS provided a tool for the CLEAN researchers in the following aspects: scientific process management, data analysis, integration and visualisation, data transfer and scheduling through specialised database systems and retrieval techniques, storage technology, and efficient data access. Without a management of the large amount of legacy industry data to be considered in the CLEAN project, the project would have never been completed in a 3-years term and the efforts needed by each project partner to fulfil their planned work would have exceeded any expectation.

It is anticipated that the DMS architecture and the developed data management tools will be used in further scientific projects of similar complexity and data sharing.

4.2 Properties of Reservoir and Cap Rocks and their Potential Alteration

For safe CO_2 injection within the framework of EGR it is required to improve the understanding about the sealing properties of the cap rock (mainly Zechstein halite) as well as the properties of the clastic reservoir sediments (mainly Rotliegend sandstones, siltstones and mudstones) by application of different analytical approaches with the focus on potential fluid-rock interactions at depth. Therefore, at first a baseline characterisation was performed comprising:

- Hydromechanical features of the reservoir and cap rock, including primary stress states and hydromechanical properties (e.g. poroelastic properties, rock strength),

- Sedimentological (litho- and depofacies) features, the mineralogical/geochemical rock composition and the evaluation of the pore space-mineral surface boundaries as well as the parameterisation of observed rock parameters of the variably altered Rotliegend reservoir sediments,
- Petrophysical features (e.g. rock porosity, permeability, irreducible water saturation and residual gas saturation) of the Rotliegend reservoir sandstones.

Second, laboratory experiments were conducted to verify the expected fluid-rock interactions in the Rotliegend sandstones under reservoir conditions (T ~120 °C, p20 MPa). Prior and after the experiments the petrophysical and mineralogical/geochemical composition of the rock samples were analysed. Also the changes in the brine composition were recorded. Thus a number of data sets of Rotliegend rocks were available, enabling geo-/hydrochemical and mineralogical modelling of potential CO_2-induced fluid-rock reactions.

4.2.1 Lithological Baseline Characteristics and Fluid-Rock Reactions in Rotliegend Reservoir Sandstones

The Altmark Rotliegend sediments comprise laminated claystones, heterolithic silty sandstones and fine- to medium-grained structureless and cross-bedded sandstones. Sedimentary structures, grain size, grain sorting and mineral content (e.g. hematite, early diagenetic pore-filling carbonate and sulphate cements) indicate deposition at semi-arid conditions on a continental, sebhka/playa-like flood plain, with braided-river systems and subordinate amounts of aeolian dunes and playa lakes. Thus two major depositional sediment types of (a) aeolian (dunes, dry sandflats) and (b) aquatic (wet to damp sandflats, fluvial channel fills) origin are preserved (Fig. 4.4). Thereby, in contrast to the red aquatic dominated sediments, greyish aeolian sandstones exhibit moderate to low contents of (matrix and coating) bounded clay minerals and early diagenetic (anhydrite, calcite) cements, but well pronounced grain size distribution and sorting, resulting in enhanced porosity and permeability (Fig. 4.5).

Remnants of red sandstone in the bleached rocks and vice versa suggest that bleaching is an alteration process affecting the pristine red bed sandstones and modifying them into bleached sandstones. Such bleaching/alteration of red sediments is commonly known and explained by interactions of (CO_2 bearing) fluids with red sandstones, in the presence of e.g. H_2S, hydrocarbons and/or organic acids (e.g. Haszeldine et al. 2005).

These fluid-rock reactions are reflected by distinct complex mineral behaviour and variations in geochemical and petrophysical composition (e.g. porosity, permeability; Fig. 4.5) in the red and the greyish rocks. In the pristine red bed rocks, carbonate and anhydrite cement dissolution is commonly restricted to cleavage planes, forming triangular etch pits in almost compositional homogeneous carbonates (Fig. 4.6a, b). Similar dissolution pattern in carbonate minerals were induced by experimental runs (e.g. Pérez-Garrido et al. 2009). In the greyish

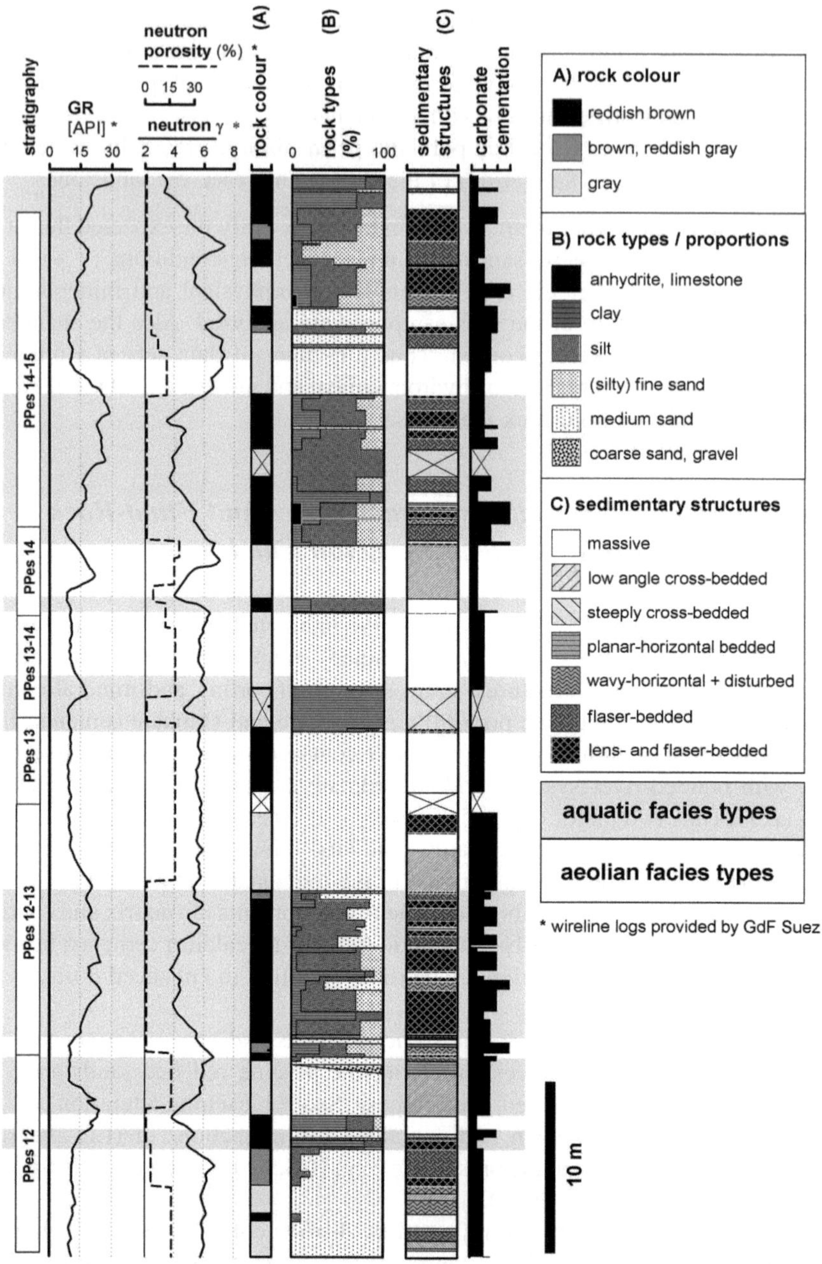

Fig. 4.4 Lithological profile of an Altmark well site with intervals of aeolian (*light grey*) and aquatic (*dark grey*) dominated sandstone sequences

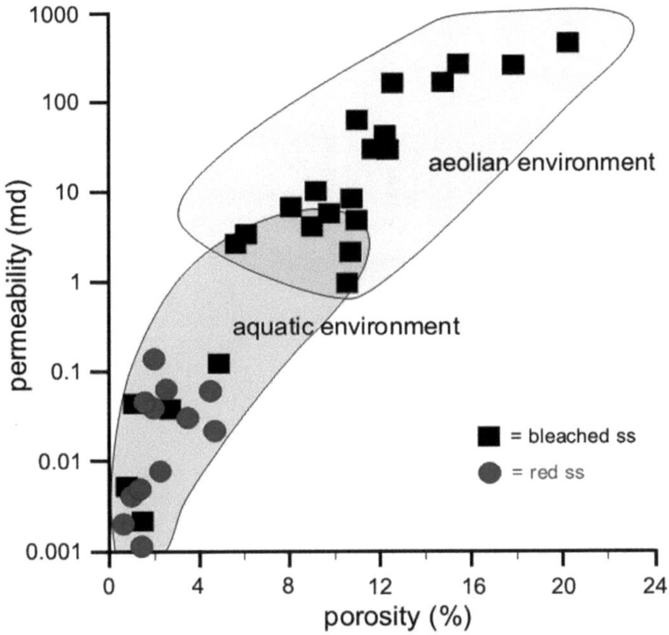

Fig. 4.5 Correlation of porosity and permeability with facies types and rock colours. Aeolian-type sandstones (*squares*) exhibit highest porosity and permeability and are preferentially bleached/altered by diagenetic fluid-rock interaction

Altmark rocks, carbonate and anhydrite display rough, irregular, heterogeneous surfaces, caused by corrosion processes and/or slightly distinct carbonate composition (Fig. 4.6c, d).

The presence of two carbonate generations within the cements, a dominating almost pure calcite phase (MnO <2 wt.%) and only minor portions of Mn-enriched calcite (MnO ~2–6 wt.%) is sustained by microprobe investigations. In both carbonate types the Fe and Mg content is <2 wt.%. These Mn enriched calcites form small (≤ ~6 μm) rounded elongated and eu- to subhedral, triangular to rhombohedral crystals within the almost pure calcite, suggesting precipitation in former dissolution holes and etch pits (Fig. 4.6d). Besides differences in carbonate (and anhydrite) design also the occurrence of clay minerals differs in red bed and bleached sandstones. In red bed sandstones close intergrowths of illitic/chloritic minerals in clast coatings are preserved, whereas coating transformation into discrete chlorite and illite aggregates is almost absent (Fig. 4.7a, b). Thereby, chlorites are characterised by high Fe, but low Si content. In the bleached rocks, intergrowths of illitic/chloritic material in the coatings are rare. Instead, clast-rimming to pore-bridging discrete chlorite and illite is frequent (Fig. 4.7c, d). The honeycomb texture of the chlorites points to a transformation of smectitic-corrensitic precursor minerals into Mg-rich, Si-poor chlorite during burial diagenesis (Hillier 1994). The presence of very thin fibres of illite in these honeycomb

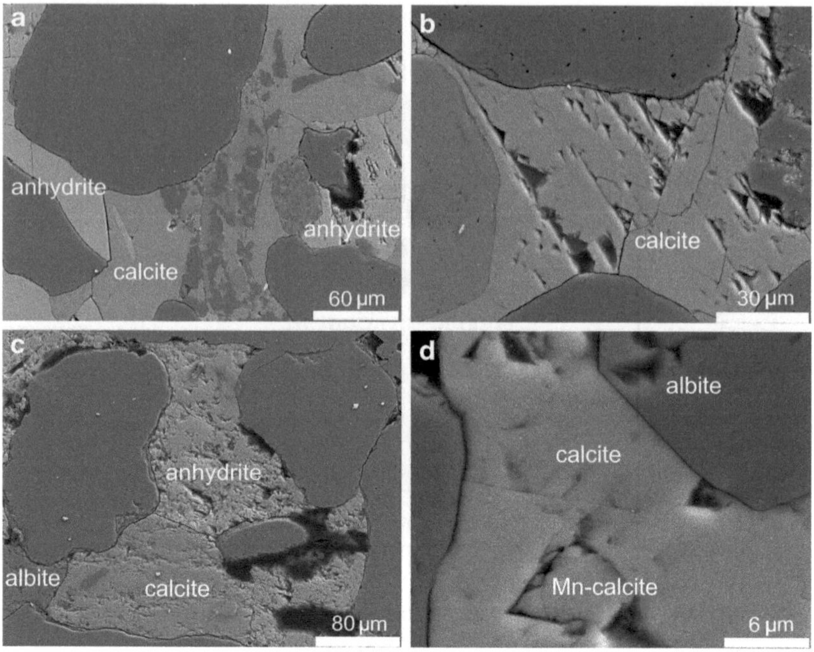

Fig. 4.6 Back scattered images of scanning electron microscopy showing varying rates of dissolution in pore-filling anhydrite and calcite cements. (**a**, **b**) Homogeneous mineral compositions with only weak dissolution of the cements in red bed sandstones. (**c**, **d**) Dissolution pattern of cements in bleached rocks are more pronounced and along with abundant (Mn-poor) calcite, also small (maybe neo-precipitated, Mn-rich) calcites are present

chlorites, on anhydrite cleavage planes and surfaces as well as over- and intergrowths with authigenic albite indicate some coeval formation of these mineral phases. Elsewhere eu- to subhedral authigenic barite and REE enriched carbonate ("synchysite"; La, Ce, Nd, and Sm oxides >50 wt.%) is over- and intergrown by chlorite. In the altered sandstones, chlorite content is related with a depletion in carbonate and anhydrite cements and porosity as well as permeability enhancement. Therefore, most probably the transformation of the illitic-chloritic coatings into discrete monomineralic phases occurred almost simultaneously with the formation of further mineral phases (e.g. albite, barite, synchysite) and the dissolution of carbonate and anhydrite cements, resulting in increased porosity and permeability in the altered greyish rocks.

Thus, the inherent enhanced pore connectivity in the bleached rocks enables high water-rock ratios during interactions resulting in more complex mineral dissolution and precipitation processes. This is in contrast to the low porosity red bed sandstones. Here, commonly small isolated pores occur almost without any contact. Thus, water-ratios are low and mineral transformation is only limited to a somewhat "semi-closed" system. This highlights the importance of physico-chemical micro-environments during diagenesis in the Altmark. The relevance of

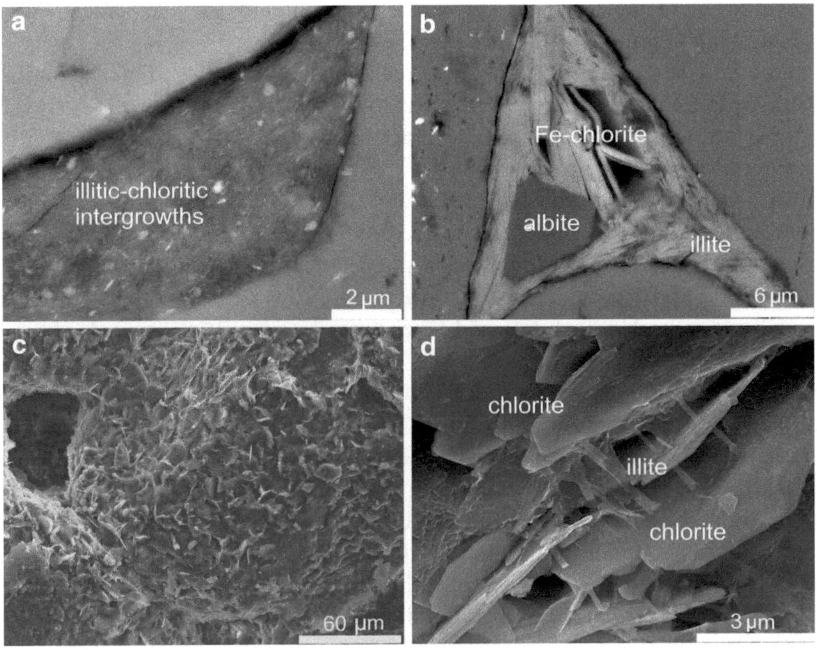

Fig. 4.7 Back scattered images of scanning electron microscopy showing distinct morphological and compositional types of illite and chlorite. (**a**) Close intergrowths of illitic and chloritic minerals along with small grains of Fe and Ti oxides in clast rimming coatings in the red bed sandstones. (**b**) In these rocks small, commonly isolated pores are filled by dense illitic aggregates along the grains (*grey*) and small Fe-rich chlorites (*light grey*) towards the pore centre. (**c**) In the bleached rocks chlorite forms honeycomb textures almost totally covering the clasts. (**d**) These Fe-Mg chlorites exhibit eu- to subhedral habits with small illite fibres

such micro-environments on mineral dissolution and close adjacent mineral formation of similar or distinct mineral species is commonly known (Boles and Johnson 1984; Kirnbauer 1997; Pudlo and Gaupp 2010). In the Altmark, this is most strikingly exposed by the habit and composition of the different chlorite types and the joint presence of Mn-poor and Mn-enriched calcites.

The preservation of initially cement filled pore space and large pore sizes implies that these reactions proceeded during deeper burial, accompanied by chlorite formation. According to their diagenetic facies these altered bleached sandstones can be classified as C (=chlorite) type facies, which typically exhibit opportune reservoir qualities, whereas the red bed sandstones are of H (= hematite-bearing) and SB (=sebhka/playa) type, due to their strong carbonate and anhydrite cementation, causing only low-grade reservoir qualities (Gaupp 1996).

Thus, bleaching of red bed sandstones into greyish rocks is accompanied by distinct mineral dissolution and precipitation reactions. Most probably these processes were induced/followed by hydrocarbon migration. This is deduced from the presence of different hydrocarbon species in red and greyish sandstones and bitumina, which is most frequent in the altered rocks.

Thereby, a sequence of fluid-rock reactions is established in the Altmark sandstones: (1) rock bleaching, (2) fibrous to meshwork-like illite and honeycomb chlorite formation, (3) hydrocarbon migration, not excluding an almost contemporary effect of these processes.

Most probably these fluid-rock reactions occurred during burial diagenesis in Triassic/Jurassic times, when Carboniferous sourced fluids passed Rotliegend volcanic rocks along (re-activated) faults and got enriched with a set of elements (e.g. REEs). Such timing is also assumed by Hartmann (1997) based on geochemical and mineralogical data and Fischer et al. (submitted) from Ar-Ar-dating of fault-related K-feldspars.

The overall consequence of these investigations is that strongly cemented red bed sandstones preserved only minor porosity and permeability for EGR processes. In contrast, bleached greyish rocks reveal high, most favourable poro-perm features. These differences are mainly caused by grain size and grain sorting (reflecting different lithofacies deposition) enabling complex fluid-rock interaction of mineral precipitation and mineral dissolution during burial diagenesis. The resulting different reservoir types will react at variable rates and modes (e.g. CO_2 trapping mechanisms) to potential fluid injection during EGR and CO_2 storage projects and will be of crucial importance for reservoir injectivity and storage capacity.

4.2.2 Experimental Determination of Fluid Transport Properties and CO_2-Induced Alteration of the Reservoir Rocks

Studies of Lutz et al. (1975) and Reitenbach et al. (2007) suggest that a common attribute of Rotliegend reservoir sandstones in the North Sea and North German area is the high water binding capacity. This feature is most likely due to their clay content, present as clast rimming coatings of mainly illitic to chloritic composition, pore-lining chlorite and pore-lining to pore-filling illite, causing high connate water content and very limited mobility of water and gas phase. Moreover, their presence is one important attribute in controlling rock porosity and permeability and thus fluid migration. Therefore, a special objective was to study and evaluate CO_2-fluid-rock interactions and their effects on fluid flow transport properties as well as on the injectivity of wells in order to evaluate the suitability of the Rotliegend sandstone formations in the Altmark for CO_2 injection.

Two consecutive series of Routine and Special Core Analysis (RCA, SCAL[1]) measurements prior and after laboratory batch runs have been performed. These

[1] Routine core analysis (RCA) generally includes porosity, grain density, fluid saturation and a lithologic description. Special core analysis (SCAL) generally includes relative permeability, wettability and capillary pressure measurements.

Table 4.1 Scope of routine and special core measurements performed. S_w = water saturation; S_{wi} = irreducible water saturation; S_{gr} = residual gas saturation

First experimental series	Second experimental series
Porosity and absolute permeability	Core flooding with CO_2 saturated brine
Core flooding	NMR at $S_w = 1$
NMR at $S_w = 1$	Specific water permeability
Specific water permeability	Capillary pressure functions
Capillary pressure functions	NMR at S_{wI}
NMR at S_{wI}	End-point gas permeability
End-point gas permeability	Relative permeability functions
Imbibition (re-flooding) → S_{gr}	Imbibition (re-flooding) → S_{gr}
Relative permeability functions	Absolute permeability and porosity

investigations comprise the determination of porosity, gas and water permeability, end-point effective gas permeability, capillary pressure functions, irreducible water and residual gas saturations, and Nuclear Magnetic Resonance (NMR) transversal relaxation time T2 distributions. A number of 56 core plugs have been routinely characterised and 17 core plugs selected for SCAL characterisation (Table 4.1).

In the first base series the measurements were performed at the original (unaltered by laboratory batch runs) core plugs by using synthetic reservoir brine and nitrogen gas. In the second series these properties were determined after plug ageing by carbonized synthetic formation brine at representative reservoir temperature and pressure conditions (20 MPa, 125 °C) to mimic the effects of CO_2 injection on the reservoir rocks. Such effects might comprise permeability alteration caused by the dissolution of pore-filling mineral cements by carbonated brine and pore-throat plugging by fines of mainly clay minerals. In closed loop flooding experiments both, increase of permeability due to dissolution of carbonates and its decrease due to pore throats plugging by illite and kaolinite fines was observed (Ross 1982; Bowker and Shuler 1991; Sayegh et al. 1990).

The synthetic formation brine used in the Clausthal experiments contained 13.71 g L^{-1} KCl, 179.13 g L^{-1} NaCl and 159.39 g L^{-1} $CaCl_2$. The static ageing was performed during 4–6 weeks lasting experiments in an autoclave. In reference, relative permeability measurements and dynamic ageing was performed in a pressure cell by closed-loop flooding of a composite core with the CO_2-saturated brine at reservoir conditions during 3 weeks. These petrophysical analyses confirmed substantial relations of porosity and permeability in regard to lithofacies types given before (Fig. 4.5). This lithofacies control on rock properties was also maintained on other fluid transport properties, like the stress dependency of permeability. This stress related attribute is much more pronounced in the low permeable aquatic facies, than in the high permeable aeolian sandstones (Fig. 4.8).

Mineral characterisation of the frequently used aeolian, dry sandflat lithotype was determined by means of X-ray diffraction analysis applying the Rietveld method, confirming significant amounts of potentially CO_2 reactive minerals like feldspar, chlorite, calcite, dolomite and anhydrite (Table 4.2). Thus, this mineral content in the sample can release and incorporate ions determined in the brine after

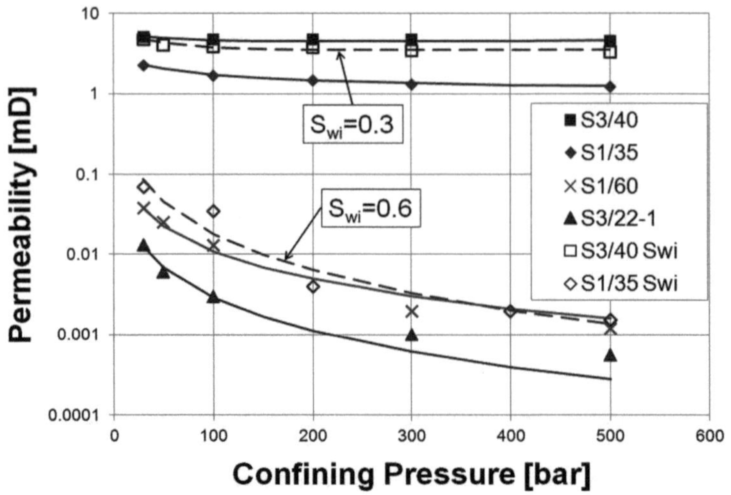

Fig. 4.8 Stress dependency of absolute and end-points effective gas permeability for aquatic lithofacies types

Table 4.2 Minerals identified by X-ray diffraction analysis in core plugs used for carbonized brine flooding

Plug ID	Illite + mica [%]	Chlorite [%]	Quartz [%]	K Feldspar [%]	Plagioclase [%]	Calcite [%]	Dolomite [%]	Anhydrite [%]	Total [%]
S1/15-1	3.0	2.8	74.6	6.2	7.4	4.1	1.0	0.9	100
S1/15-2	3.8	2.8	73.3	6.9	7.5	3.6	1.0	1.1	100

the ageing runs. Table 4.3 suggests complex dissolution and precipitation processes caused by CO_2-rock-reactions.

Primarily measured capillary pressures of aeolian and fluvial sandstone lithotypes show clear differences (Fig. 4.9), which can also clearly be discerned from their Leverett J-functions (Fig. 4.10), indicating changes in the pore space structure of the test cores after the ageing (Fig. 4.10). This effect can be attributed to the alteration of the pore size distribution and porosity-permeability relationship.

NMR T2 spectra of aged samples exhibit a pronounced shift towards larger relaxation times (Fig. 4.11). This suggests an increase in water mobility caused by an enhancement in porosity.

However the increase in brine acidity can reduce these effects of aqueous phase mobilisation. This is evident from the NMR T2 distribution in Fig. 4.11, when corrected to pH = 4, as estimated for the reservoir conditions of p = 20 MPa and T = 125 °C. In combination of these effects the specific water permeability of sandstones after ageing slightly decreases (Fig. 4.12).

Irreducible water saturation was determined by draining the brine saturated samples with nitrogen in capillary pressure cells up to a capillary pressure of 1.8 MPa. The irreducible water saturation of the unaltered plugs exponentially

Table 4.3 Ion concentrations in carbonized brine before and after ageing of plug sets in autoclave

Ion ID	Contents before ageing, [mg/L]	Contents after ageing, [mg/L]	
		First set of plugs	Second set of plugs
Sulfate, SO_4^{-}	<20	198.0	400.0
Hydrogencarbonate, HCO_3^{-}	<6	567.0	787.0
Iron, Fe^{++}	<0.05	98.5	41.9
Magnesium, Mg^{++}	10.1	54.4	75.0
Calcium, Ca^{++}	42,190	26,400	25,520

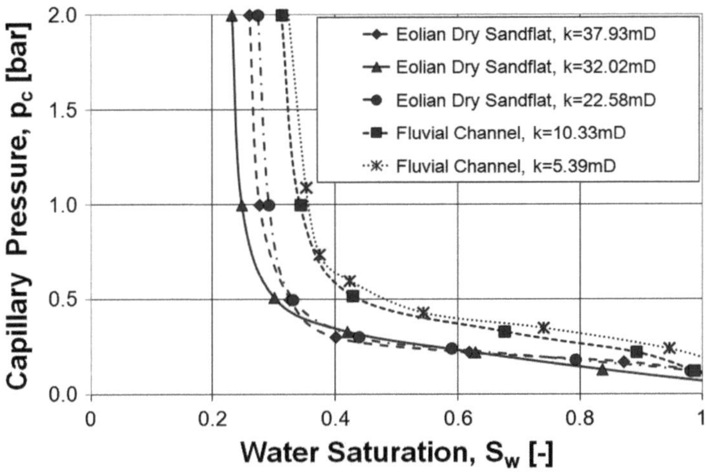

Fig. 4.9 Capillary pressure functions of aeolian and fluvial lithofacies

increases with the decreasing permeability (Fig. 4.13). This trend is more pronounced for medium to high permeability aeolian dry sandflat facies and less for the aeolian dune facies.

The residual nitrogen gas saturation was determined by re-flooding the drained plugs in a Hassler cell with synthetic formation brine at pressure gradients typical for distant reservoir areas. After the passing through the core brine volumes equal to 5–10 pore volumes of the core flooded, the residual gas saturation stabilised at a level of 40–55 %. This range of residual gas saturation is very common for different sandstones studied by other authors (Katz et al. 1966). In application to the studied Altmark sandstones this verifies the essential potential to immobilise CO_2 in water re-flooded areas of the reservoir.

The drainage and re-flooding experiments were repeated with the same plugs after their ageing during the batch experiments. Thereby, the induced rock alteration is reflected by a pronounced increase in irreducible water saturation (Fig. 4.13) and a decrease in residual gas saturation. This effect can possibly be related to the alteration of the pore cement minerals due to CO_2-rock interactions

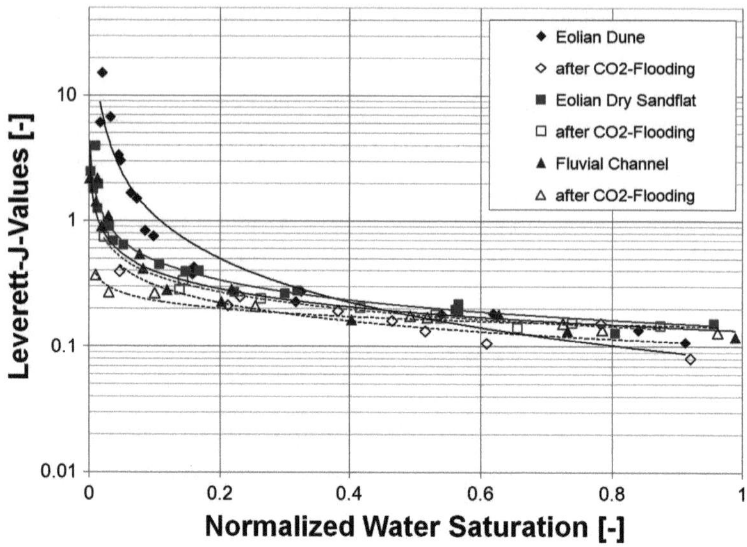

Fig. 4.10 Leverett-J-functions of main lithofacies before and after ageing in carbonized brine

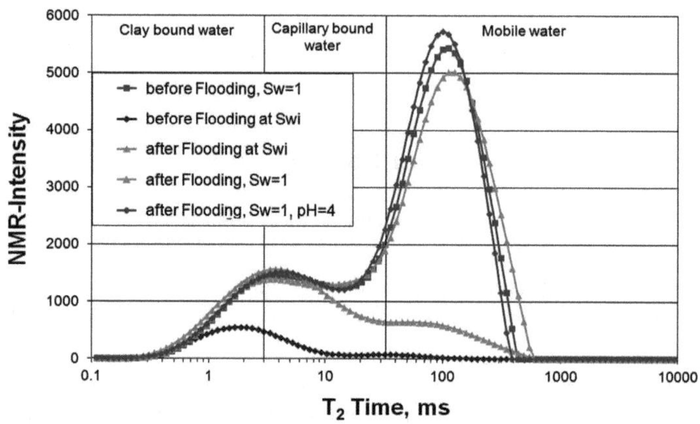

Fig. 4.11 Effect of ageing in carbonized brine on T2 distributions of a medium permeable fluvial channel sandstone

causing an increase in wettability and water binding capacity. Thus the CO_2 injection could reduce the mobility of the aqueous phase. This has crucial importance for forseen CO_2 injection, because injectivity of the wells will strongly depend on the immobile water phase saturation and mobility of the gas phase in the near wellbore area.

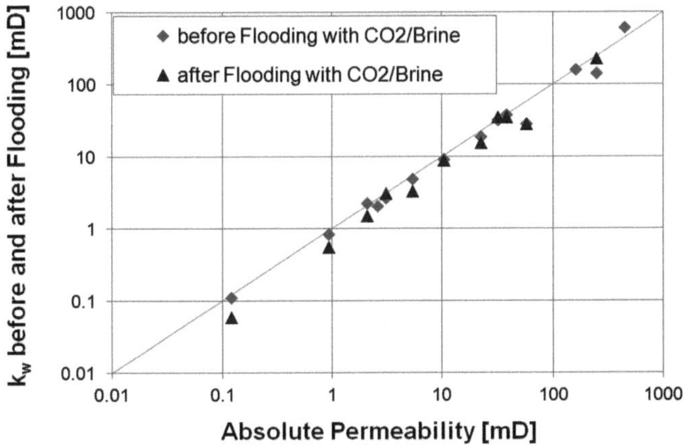

Fig. 4.12 Effect of ageing in carbonized brine on specific water permeability

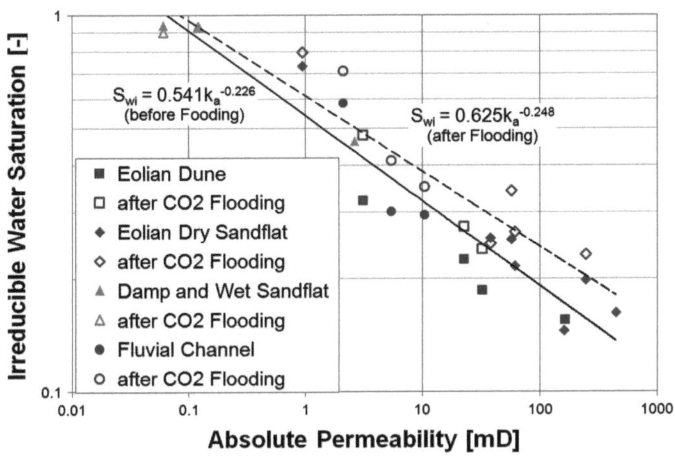

Fig. 4.13 Irreducible water saturation prior and after ageing in carbonized brine

4.2.3 Determination of 3D Primary Stresses and Hydromechanical Rock Properties

The determination of the 3D in situ stresses at the time of coring (=primary stresses) was performed using the RACOS® method (Braun 2009). It is based on direction-related analysis of the ultrasonic elastic wave propagation in loaded rock samples. Results are obtained on the loading-dependent propagation velocities of compressional and shear waves to derive information about the rock structure.

The properties presented are determined on five sections of rock cores retrieved from the wellbores S1, S3 and S76 located in the Altensalzwedel area (Hou et al. 2009a). Table 4.4 provides descriptions and hydro-mechanical properties of the

Table 4.4 Hydromechanical properties of core material from the wellbores Altensalzwedel S1, S3 and S76. Density, porosity, dynamic Young's moduli, dynamic Poisson's ratios and Biot coefficients as obtained in Braun (2009).

Wellbore		S1		S3		S76
Rock type		*Siltstone*	*Claystone*	*Siltstone*	*Sandstone*	*Anhydrite*
In situ depth (m)		3,429	3,478	3,487	3,513	3,298
Density (g/cm^3)		2.69	2.72	2.71	2.46	2.94
Porosity (%)		0.7	2.5	1.1	7.0	0.1
E (GPa)	RACOS® (dynamic)	70.5	55.8	67.5	53.6	83.7
	Modified (static)	28.2	22.1	26.9	21.2	33.6
v (−)		0.186	0.214	0.195	0.129	0.274
a (−)	RACOS® (mean)	0.172	0.358	0.210	0.645	0.082
	Modified	0.172	0.358	0.210	0.645	0.000

investigated samples. Also it lists density, porosity, dynamic Young's moduli, dynamic Poisson's ratios and Biot coefficients obtained in Braun (2009).

Density and porosity are measured under average in-situ loading. Among these, the density and the dynamic Young's modulus exhibit rather large values. This may be attributed to the small rectangular block samples prepared from relatively homogeneous sections of cores, and might not truly represent the in-situ conditions where the formation contains an unknown number of discontinuities and fractures with unknown directions. For the numerical modelling, the static instead of the dynamic Young's modulus is used. The static Young's moduli from Table 4.4 are converted with the following correlation function (Wang 2000):

$$E_{static}(GPa) = 0.4145\ E_{dynamic}\ (GPa) - 1.0593 \tag{4.1}$$

RACOS® measures the pore pressure effectiveness ω_{ij}, which is direction-dependent (Braun 2009). Since the differences between the three principal pore pressure effectivenesses, especially in permeable rock formations, are relatively small and there are no numerical codes which take ω_{ij} into account, the direction-independent Biot's coefficient is proposed, which has the average value of the three principal pore pressure effectivenesses ω_1, ω_2, ω_3, as follows:

$$\alpha = \frac{\omega_1 + \omega_2 + \omega_3}{3} \tag{4.2}$$

The effective in-situ stresses at the time of coring directly measured by RACOS® specialised analyses are summarised in Table 4.5. Six effective stress components are calculated from three principal effective primary stresses (σ'_{10}, σ'_{20}, σ'_{30}). Shear stress components exhibit almost negligible values, revealing that the primary stresses σ'_v, σ'_H, σ'_h, could be seen as principal stresses. Furthermore, it is more convenient to use σ'_v, σ'_H, σ'_h than to use σ'_{10}, σ'_{20}, σ'_{30} e.g. for coupled numerical modelling.

Especially for the location of wellbore S76 (outside of the Altensalzwedel area), the determined direction of the maximum effective horizontal stress (σ'_H) is

Table 4.5 Magnitude, azimuth and dip of principal effective primary stresses and calculated maximum, minimum horizontal and vertical stresses and shear stresses of core samples from wells S1, S3 and S76

Wellbore		S1		S3		S76
Rock type		*Siltstone*	*Claystone*	*Siltstone*	*Sandstone*	*Anhydrite*
σ'_{10} (max) (MPa)		70.3	60.3	69.3	53.3	77.2
σ'_{20} (inter)		68.6	40.6	63.5	48.2	70.9
σ'_{30} (min)		54.2	29.4	56.0	37.4	63.8
Azimuth of $\sigma'_{10/20/30}$ in °	max	134	74	85	237	337
	inter	231	213	233	58	242
	min	321	304	331	327	67
Dip of $\sigma'_{10/20/30}$ in °	max	77	82	59	55	1
	inter	2	6	27	35	78
	min	13	5	14	0	12
Azimuth of $\sigma'_{H/h}$ in °	σ'_H	51	35	65	57	157
	σ'_h	141	125	155	147	67
Magnitude of $\sigma'_{H/h/v}$ (MPa)	σ'_H	68.6	40.8	64.7	49.9	77.2
	σ'_h	55.0	29.7	56.8	37.4	64.1
	σ'_v	69.6	59.8	67.3	51.6	70.6
Magnitude of $\tau'_{vH/vh/hH}$ (MPa)	τ'_{vH}	0.0	2.0	1.0	−2.4	−0.1
	τ'_{vh}	3.4	2.8	2.7	0.4	−1.4
	τ'_{hH}	0.0	−0.3	−0.7	0.6	−0.3

NNW-SSE (azimuth 157°). This corresponds approximately to the direction of the marked main fault zone. The smallest in-situ loading component is the minimum effective horizontal stress (σ'_h). This stress condition corresponds to that of a strike-slip fault. In all investigated sections of the Altensalzwedel wellbores S1 and S2, the maximum effective horizontal stresses act approximately NE-SW. This direction corresponds to further fault zones of smaller extent than the main fault zones in the Altensalzwedel area recognizable from structural maps. The stress orientations determined for this section of the reservoir correspond to those of a normal fault.

4.2.4 *Parameter Study Estimating Potential Mineral Alteration in the Reservoir*

A crucial constraint for simulations with regard to reservoir rock alterations is the availability of required databases for different codes applied. Available and accessible thermodynamic data for geochemical reservoir simulations were compiled (Moog and Mönig 2009) and databases created for the codes PHREEQC (Parkhurst and Appelo 1999) and CHEMAPP (Petersen and Hack 2007a, b). These databases especially consider temperature corrections of the thermodynamic equilibrium constants as well as the activity constants after Pitzer et al. (1984) for highly saline solutions to represent the conditions of the Altmark reservoir.

Despite the fact that mineralogical-petrographical analyses of the rocks yielded some parameters usable for reaction kinetics (BET-analyses, qualitative and quantitative characterisation by image analyses of pore exposed mineral phases), reaction rates cannot be derived with the available data. In this situation, numerical simulations for the depleted gas reservoir can only yield simplified qualitative information and trends, but cannot be used for substantiated, quantitative predictions of geochemical reactions during CO_2 geological storage.

A number of approaches exist for the description of CO_2 solubility in saline solutions. For the geochemical simulations addressed in the following, the amount of dissolved CO_2 was determined using the approach of Duan et al. (2006) in an initial NaCl–$CaCl_2$–KCl (molar ratio 16.7:7.18:1) solution representing a simplified brine composition as typical for the Altmark. CO_2 solubility was determined as a function of CO_2 partial pressure (p), temperature (T), and ionic strength.

A total of 210 simulations were performed for a seven-step-variation of CO_2 partial pressures between 0.1 and 80.0 MPa, five temperature steps between 1 °C and 150 °C and six dilution steps of the initial NaCl–$CaCl_2$–KCl solution of 0 (pure water) to 2.8 mol NaCl per kg of water.

As no calculations that include kinetic considerations can be performed to date for the studied area, a simplified incorporation of CO_2 into the calculations was used, with five times as much CO_2 as can be dissolved at the respective thermodynamic equilibrium conditions in 1 kg of water in each calculation. Considering that small amounts of water from the residual water saturation will dissolve in the CO_2 phase (dry-out-effects) or that a progressing CO_2 phase will displace a continuous aqueous phase (if present) the assumption of only a limited amount of CO_2 being available for the reaction is reasonable for most parts of the Altmark gas field. On the other hand, the amount of CO_2 being available for reactions might be higher than the dissolved amount at the respective thermodynamic equilibrium conditions as the pore fluid might be replaced by flow processes, or additional CO_2 might dissolve in the pore fluid while CO_2-consuming reactions take place.

For the simulations a representative, unaltered primary rock composition was defined of quartz, potassium-feldspar, albite, chamosite, illite, hematite, calcite and anhydrite as found in rocks of the Altmark. The molar amounts are normalised to a pore volume of 1 L. Precipitation of additional minerals was not considered, although the simulations yielded a positive saturation index for a number of silicates (e.g. kaolinite), indicating their potential for precipitation. Due to lacking thermodynamic data sets even some described discrete mineral phases (e.g. Fe-Mg-chlorite, anatase, barite, Mn-rich calcite) could not be included in the simulations. Close mineral intergrowths (forming clast rimming coatings composed of varying amounts of illite and chlorite) and compositional highly variable lithoclasts are only rudimentary regarded, assuming most realistic volume ratios of observed and analysed mineral types. Although the content of these mineral phases is only minor, their dissolution can strongly affect rock reactivity by the release of cations (Ca, Mg, Fe, Mn) favouring carbonate (neo-) precipitation (e.g. Shiraki and Dunn 2000; Kirste et al. 2004; May 2005). Considering the given restrictions the simulations provide a first rough orientation regarding effects of the CO_2 partial pressure, temperature and salinity on geochemical reactions.

Figure 4.14 shows the general tendency of increasing precipitation of calcite with increasing partial pressure of CO_2 (p). However, with increasing temperature (T) this tendency decreases. Increasing the ionic strength of the initial

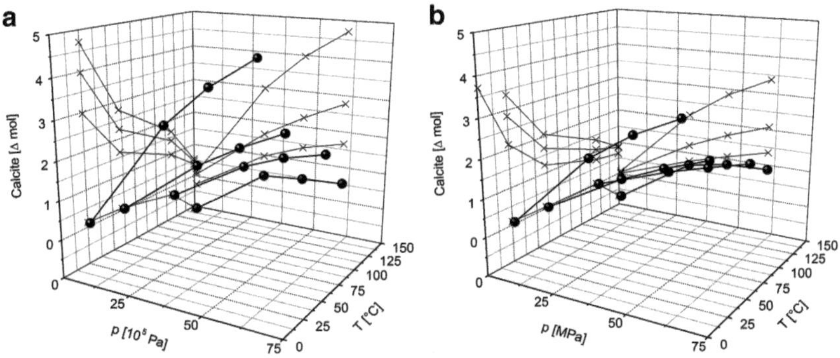

Fig. 4.14 Precipitation (positive values) of calcite (in mol) as a function of CO_2 partial pressure (p), temperature (T) and ionic strength of the initial NaCl–$CaCl_2$–KCl solution with (**a**) 1.9 and (**b**) 3.7. In both figures spheres indicate the mineral amount within the calculated pT-range (*bold lines* connect *spheres* at equal temperature) and projections (*crosses* connected with *thin lines*) are provided for better understanding the temperature and pressure effect on the calculated amount

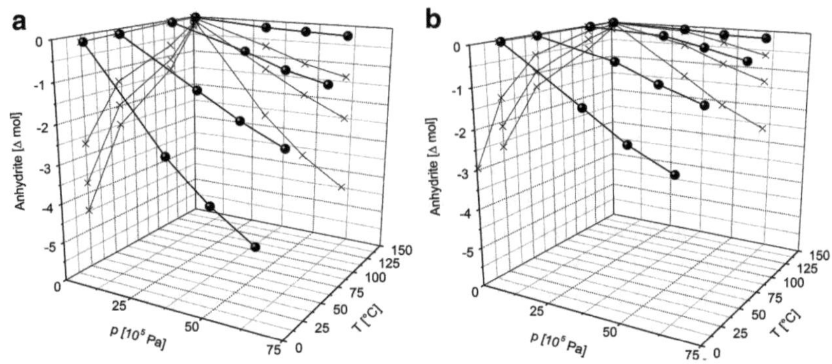

Fig. 4.15 Dissolution (negative values) of anhydrite (in mol) as a function of CO_2 partial pressure (p), temperature (T) and ionic strength of the initial NaCl–$CaCl_2$–KCl solution with (**a**) 1.9 and (**b**) 3.7. In both figures spheres indicate the mineral amount within the calculated pT-range (*bold lines* connect *spheres* at equal temperature) and projections (*crosses* connected with *thin lines*) are provided for better understanding the temperature and pressure effect on the calculated amount

NaCl–$CaCl_2$–KCl solution shows no clear tendency. With an ionic strength of 1.9 (Fig. 4.14A) at 25 °C and a CO_2 partial pressure of 6.0 MPa precipitation of more than 4.5 mol calcite is calculated, and at 150 °C a precipitation of only about 0.5 mol is calculated. In the calculations with an ionic strength of 3.7 (Fig. 4.14b) at 25 °C and a CO_2 partial pressure of 6.0 MPa less than 3.5 mol calcite precipitate, but at 150 °C nearly 1 mol calcite precipitates. Even further, with only 0.1 MPa of CO_2 partial pressure dissolution of existing calcite is calculated at higher temperatures around 120 °C and 150 °C.

Figure 4.15 shows the general tendency of increasing dissolution of anhydrite with increasing partial pressure of CO_2 (p). With increasing temperature (T) this tendency decreases. For anhydrite dissolution an increasing ionic strength of the initial NaCl–$CaCl_2$–KCl solution (1.9 for Fig. 4.15a, 3.7 for Fig. 4.15b) shows a general trend of decreasing anhydrite dissolution for all calculated temperatures.

Amongst other features the adaptation of these calculations to more realistic scenarios for the Altmark gas field will require the consideration of reaction kinetics and an assessment of the amounts of minerals which are actually accessible for reaction as wells as a detailed compilation of minerals determined by the petrographic investigations.

4.3 State of Stress in the Overburden and Perspectives for Long-Term Monitoring of Possible Changes

The overburden of the Altmark gas reservoir is structured as a multi-layered sequence of salt layers of Zechstein age and of younger marine or terrestrial sediments from Triassic to Quaternary. The Zechstein formations (including the Zechstein salt) constitute the immediate geological barrier for fluids contained in the subsalt porous reservoir of clastic sedimentary rocks of Permocarboniferous age. This sealing cap rock layer needs to ensure as well the permanent storage of injected CO_2 within the reservoir, besides other trapping mechanisms. This sealing quality must be evaluated. However, the quality of the cap rock and overburden integrity depends not only on the gas permeability of the rocks, but also on the sealing performance of natural tectonic fractures and faults within the rock sequence. Natural joints or faults in the cap rock sequence and its overburden are the most decisive pathways for fluids. A necessary consequence for any future EGR or CO_2 storage project is the monitoring of the state of stress as inevitable part of the programme. Such a stress monitoring can determine changes in the cap rock due to the injection process or prove insignificant consequences for the sealing quality of the overlying geological formations.

The technical feasibility of monitoring stress developments at depths of >2–4 km show several limitations: On the one hand, the use of conventional 'hydraulic fracturing techniques' in the borehole is not permitted because of the possible fracturing that would be caused in the surroundings of a borehole ultimately increasing the permeability. That could be avoided easily with an alternative monitoring technique. On the other hand, there is currently no permanent measurement system available for these depths that might be installed under the corresponding in-situ pressure and temperature conditions and that might work sustainably.

Due to these limitations it was necessary to develop an effective stress measurement system that is adjusted to the regional stress conditions and that can be evaluated. This appropriated programme consists of (1) a summary of the knowledge about the state of stress in the Altmark and (2) construction of a test facility that will be useful to develop a suitable measuring technique in the laboratory which may be implemented successfully thereafter at the relevant depths in the steel liners of existing boreholes.

4.3.1 Methods of Drilling Data Evaluation and Lab Testing

Borehole data available within the framework of the CLEAN Project was used to evaluate the character of the regional stress field in the Altmark. The stress analysis data were used to define the required parameters of the experimental set-ups and their implementation.

The spatial distribution of stress anisotropy in the cap rocks and the overburden was deduced mainly from calliper data of about 40 boreholes as well as formation-specific gradients of essential stress components. These interpreted data were presented in a 3D GIS model. Under laboratory conditions the suitability and functionality of strain sensors were tested and verified. The relevant in-situ stress conditions are simulated with sufficient accuracy and reliability that the deformation of the "lab-model-steel-tube" corresponds to the field conditions in deep boreholes where the liner is cemented. Laboratory experiments of the simulated stress field ask for a perspective to develop a hard-inclusion stress monitoring tool that can be useful to monitor the in-situ stress in existing boreholes without damaging the cap rocks and the overburden.

4.3.2 State of Stress in the Overburden with Respect to the Natural Stress Conditions in the North German Basin

The regional large scale development of layered sequences in the overburden with the thick Zechstein salt layer resulted, with time and space, in a decoupling effect of the state of stress with three compartments in the North German Basin: (1) the suprasalt domain, (2) the Zechstein salt layer and (3) the subsalt domain. Stress orientations (e.g. S_H as maximum and S_h as minimum horizontal stress components respectively) and stress gradients or stress-magnitude changes with depth differ significantly between the Permocarboniferous subsalt and younger Triassic suprasalt formations. On the one hand, the Zechstein salt formation shows isotropic stress conditions while, on the other hand, it decouples the stress conditions between the formations above and below (Röckel and Lempp 2003; Moeck et al. 2009; Cacace et al. 2008).

4.3.3 Stress Anisotropy from Borehole Breakouts

The formation of borehole breakouts or the ovalisation of the original circular borehole cross-sections has long been regarded as an indication of distinct horizontal stress anisotropy (Bell and Gough 1979; Blümling 1986). In the numerous wells in the Altmark, borehole breakouts have been identified in the cap rocks and the overburden, thereafter categorised accordingly in order to use them as a scale for stress anisotropy analysis. It is also justified, based on the comparative investigations, that the intensity of borehole breakouts is not dependent on the

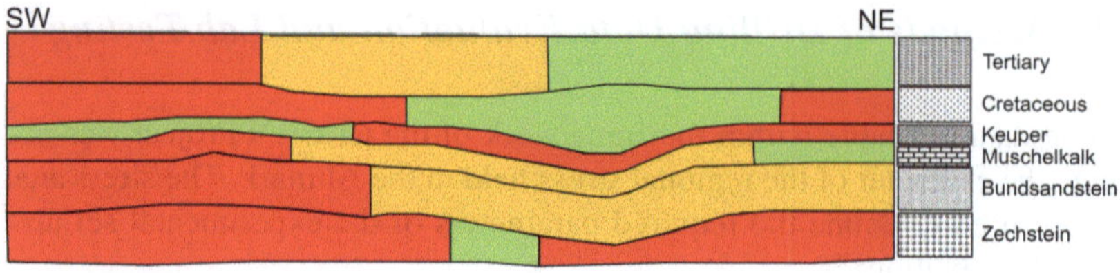

Geological Units	Low	Medium	High
	breakout category A in % of borehole length		
Tertiary	65	70	75
Cretaceous	80	85	90
Jurassic	70	75	80
Keuper	70	75	80
Muschelkalk	75	80	85
Buntsandstein	65	70	80
Zechstein	75	85	85

Fig. 4.16 SW-NE-Profile of the Altmark with relatively high, medium and low stress anisotropy between $S_{V,H}$ and S_h, derived from borehole ovalisation

deformation characteristics of the rock properties (brittle or ductile), but is a clear indication of local horizontal stress anisotropy.

For different geological formations the regional distribution of borehole breakout intensity was analysed. The spatial distribution of stress anisotropy in the cap rocks in Altmark gas field was represented in a 3D GIS model that differentiated the anisotropy semi-quantitatively in three classes of high, medium and low in layers where the formation-specific delineation of the stress anisotropy classes were determined (Fig. 4.16).

The available borehole data made it possible to determine the stress gradient and relevant stress difference in the suprasalt domain taken to represent the corresponding regional stress conditions. Thereafter, a stress difference of 9–15 % of the vertical stress component can be calculated between S_v and S_h in the suprasalt domain (Fig. 4.17). This relatively low stress difference for cap rocks (in comparison to the subsalt difference: $S_v - S_h$ of about 40 % of S_v), was used in the following laboratory experiments.

4.3.4 *Laboratory Experiments for Development of a Monitoring Device*

Based on laboratory experiments it was assessed if, on the one hand, the hard-inclusion method based on the deformation measurements (preferably in liner-steel tubes) in boreholes is feasible with sufficient accuracy (realisation of in-situ stress

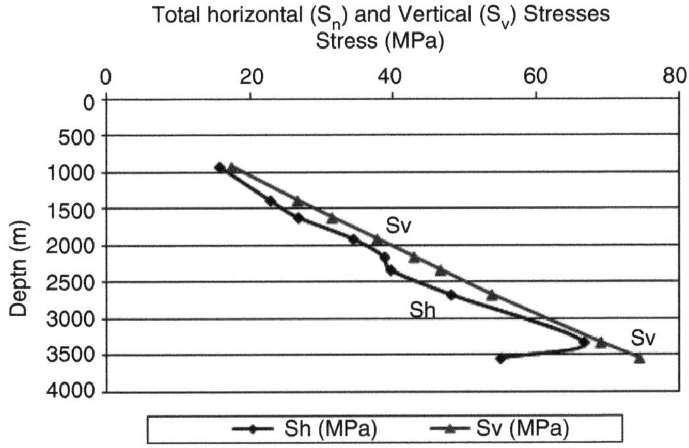

Fig. 4.17 Stress gradients of S_v and S_h in the Altmark-Region determined from available data

at depths of 2–4 km with S_v from 50 up to 100 MPa required), and, on the other hand, a high measurement resolution (min. 1/1,000 mm) can be ensured. For accurate strain and stress measurements temperature effects must be compensated.

The commonly used hard-inclusion stress measurement technique needs to be replaced with suitable strain sensors due to the limitations of the effectiveness of the hydraulic system at expected temperatures as well as pressure values at great depths. Nevertheless, the repeatability and high sensitivity of the conventional hard-inclusion pressure cells are convincing enough for stress measurements with oriented installations in steel tubes.

To realise experimentally the stress conditions, the known stress differences need to be optimised for the experimental programme that will approximate the state of stress in the Altmark (see section above), i.e. at relevant depths, respectively at corresponding pressures, a differential stress of S_v (or S_H) – S_h of 0.85 S_v to 0.91 S_v can be expected. These conditions are to be simulated in the laboratory experiments, and the consequent results on the hard-inclusion tool (that is the liner) were recorded.

Comparisons of different rock materials from the overburden lithologies (sandstone, limestone and anhydrite) show differences in the intensity of the measured deformations of the steel tube liners for the specific stress equivalents. The accuracy and reproducibility of the deformation measurements of the steel tube liners are enough to assess the developing stresses in the cap rocks when lithology and stress levels are known. By comparing the laboratory experiments with the lithology-dependent numerical model, it is possible to derive the stress from liner deformations. The principle can be implemented for any current in-situ liner geometry.

Based on the described results it can be recommended that a corresponding hard-inclusion tool for stress monitoring in different lithologies (brittle or ductile) can be implemented in cased boreholes. With the knowledge of the geomechanical properties of the cap rocks it will be possible to attain appropriate interpretation of measurement data.

4.4 Numerical Simulation of Thermo–Hydro–Mechanical–Chemical Processes during CO_2 Injection

The capabilities for EGR during CO_2 injection and the long-term safety of depleted gas reservoirs are controlled by several subsurface transport processes and geomechanical deformation effects. Among the transport processes advection of CO_2 itself and displacement of residing pore fluids, diffusion and reactive transport are frequently regarded as the most important. There is also a need to understand the long-term fate of CO_2 within a potential storage complex in order to design monitoring programmes and interpret monitoring results.

The only practical option for predicting the long-term behaviour of CO_2 in reservoirs is numerical analysis, supported by the understanding gained from the relatively short-term laboratory and field-scale experiments available. Scientific simulation based on fundamental mathematical methods is able to model the behaviour of CO_2 injected in geo-reservoirs including the interactions with the materials constituting the formation (e.g. rocks and pore fluids).

Modelling and simulation of the above mentioned thermo–hydro–mechanical–-chemical (THMC) processes as well as their mutual couplings are essential for the proper understanding of the physical and chemical processes at different length and time scales in the subsurface. This is the basis to assess the EGR and CO_2 storage capabilities of the depleted gas reservoir in the Altmark.

In order to fully characterise the geological complex under consideration, the reported modelling approach is focused on the thermodynamic flow simulation in the borehole, the analysis of the coupled THMC processes in the reservoir near-field, and the assessment of safety and EGR capabilities of the entire Altmark natural gas field during CO_2 injection. This approach comprises geomodelling (definition of the static geological model), dynamic modelling, including history matching, and scenario simulations at reservoir scale representing the full modelling and monitoring chain.

4.4.1 A Systematic for CO_2 Benchmarking

Benchmarks represent well-defined examples for process simulation under simplified conditions, validated by analytical or experimental results. Within this context, benchmarking provides the ultimate bridge between modelling and application. One of the objectives of the project was the development and implementation of a concept for complex CO_2 benchmarks, reflecting the relevant physico-chemical processes, to assure code quality (code functionality and accuracy). In order to validate project results for individual physico-chemical processes, classical benchmarks were used which are well-known from literature. For defined complex test examples based on data from application sites, a range of scientific simulators were used, such as ECLIPSE, TOUGH/FLAC, BOCO, FEFLOW and OpenGeoSys (OGS). Verification methods rely on both classic test cases and inter-code comparison (Class et al. 2009). The benchmarking strategy comprises three aspects:

- Process-based: Numerical analysis of individual and coupled processes related to CO_2 injection and storage with increasing complexity, i.e., compressible flow (H), two-phase flow (H^2), consolidation (H^2M), thermo-mechanics (TM) up to non-isothermal two-phase flow consolidation (TH^2M processes).
- Thermodynamics-based: Increasing complexity of material behaviour, i.e., from constant to highly non-linear material functions.
- Scenario-based: Development of site-specific test cases (application benchmarks).

4.4.2 Modelling Real-Gas Behaviour for Enhanced Gas Recovery Applications

For modelling a realistic behaviour of gases, it is necessary to describe the relation between temperature, pressure and volume of that gas. The mathematical formulation of this description is called equation of state (EoS), and allows determining the volume of a gas at certain pressure-temperature conditions. The knowledge of gas volume (or density respectively) is essential for modelling gas transport. Unlike liquids, the density of gases is very sensitive to pressure changes. So, fluid properties of a gas (such as viscosity, heat capacity, thermal and hydraulic conductivity, compressibility, thermal diffusivity) depend strongly on its density, and the mathematical correlations of these properties are functions of density and temperature in most cases.

4.4.2.1 Equations of State: Selection and Comparison

In this section four different, well-known equations of state are discussed and compared according to accuracy, simplicity and suitability for the required purposes. Finally, the most suitable EoS for simulating the EGR process is proposed (this was as well used for additional work). The EoS comparison included:

- A cubic equation of state presented by Peng and Robinson (1974). Cubic equations are easy to solve since they can be computed directly using Cardano's method. In general, they are tuned only by a few fitting parameters, so their potential accuracy is lower compared to more complex EoS.
- A virial EoS (Duan et al. 1992), which represents an expression in a series form of virial coefficients. These coefficients can be determined experimentally and the equation's accuracy depends on the number of terms. So, with increasing number

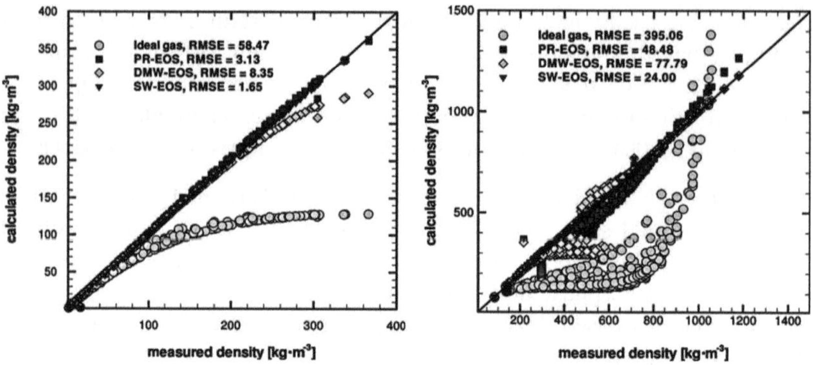

Fig. 4.18 Scatter-plot of measured and calculated density of CO_2 in the gas region (T > Tc, p < pc, *left*) and in the supercritical region (T > Tc, p > pc, *right*). Calculated densities are obtained by ideal gas law, Peng-Robinson EoS, Duan-Møller-Weare EoS and Span-Wagner EoS

of virial coefficients, the equation becomes harder to solve but its potential precision is increasing.

- A fundamental EoS known as Span-Wagner (Span and Wagner 1996): The Helmholtz free energy or Gibbs free energy can be used to determine thermodynamic fluid properties such as density. Due to a large number of fitting parameters, the potential accuracy of this correlation is very high, but the complexity of the formulation causes high costs of computing time.

To show the accuracy of these equations of state, their outcome has been compared with a total of 1,465 data sets of density measurements provided by various authors. Figure 4.18 shows scatter-plots of measured and calculated densities for CO_2 at the gas region and at the supercritical region. For comparison, ideal gas law results have been included as well. The plots show clearly that Span and Wagner's formulation provides the most accurate results among all equations presented here, which is achieved by using about 200 fitting parameters that have been tuned by a huge number of measurements. The absolute averaged deviation (AAD) from all measurements of Span and Wagner's formulation is about 1.79 % in the supercritical region and only 0.65 % in the gas region. The Duan-Møller-Weare EoS is less accurate both in the supercritical region (AAD = 8.42 %) and in the gas region (AAD = 1.59 %). Surprisingly, the simple cubic EoS by Peng and Robinson gets better results in the supercritical region (AAD = 6.27 %) than Duan-Møller-Weare EoS, but shows larger deviations in the gas region (AAD = 1.93 %).

The data presented in Fig. 4.18 and the deviations from measurements show that the Peng-Robinson EoS gives satisfying results compared to the low computational effort. Compared to Span-Wagner EoS, the results are less accurate but still tolerable for our work. Thus, Peng-Robinson's EoS was chosen for further simulations.

4.4.2.2 Model Development

The transport of gases in porous media can be described by the continuity equation. Since gases are completely miscible we consider a single phase system, and we can write:

$$\nabla(\rho v) + \frac{\partial(n\rho)}{\partial t} = \rho q_0 \tag{4.3}$$

Where ρ is density, v is Darcy velocity, n is porosity, t is time and q_0 is a source/ sink term. We consider isothermal conditions and a rigid porous medium (T = const, n = const), so the mass balance term can be developed to

$$\frac{\partial(n\rho)}{\partial t} = n\rho\chi\frac{\partial p}{\partial t} \tag{4.4}$$

where $\chi = (1/\rho)(\partial\rho/\partial p)_T$ is the isothermal compressibility of the fluid. The derivative $(\partial\rho/\partial p)_T$ may be difficult to determine since equations of state are complex functions. So, ρ is replaced in Eqs. 4.3 and 4.4 by the compressibility factor Z = (Mp/RρT) and a formulation of the continuity equation suitable for real gases is obtained:

$$\nabla\left(\frac{kp}{\eta Z}\Delta p\right) = \frac{\eta\chi p}{Z}\frac{\partial p}{\partial t} - \frac{p}{Z}q_0 \tag{4.5}$$

4.4.2.3 Benchmark Test Example

To check the validity of Eq. 4.3, a benchmark test was performed of a gas injection scenario into a homogeneous gas reservoir. The reservoir is considered to be one dimensional and axially symmetric around an injection well (Fig. 4.19).

It can be described as an annulus, where the inner radius r_0 is the borehole radius with r_0 = 1 m. The outer radius R denotes the boundary of the reservoir at a distance to the centre of R = 1,000 m. The permeability of this reservoir is $k = 1.0 \cdot 10^{-14}$ m^2, the porosity n = 0.1 is constant over time since there is no deformation of pore space due to pressure gradients. An analytical solution of this problem is given by Haefner et al. (1992).

Boundary conditions of this problem are constant injection pressure at the borehole and closed boundary at the outer rim. In order to correspond to a reservoir depth of about 4,000 m, a high initial reservoir pressure of p_i = 35 MPa was chosen, a borehole pressure of p_b = 40 MPa, and isothermal temperature conditions of T = 330 K. The working fluid in this example is pure carbon dioxide,

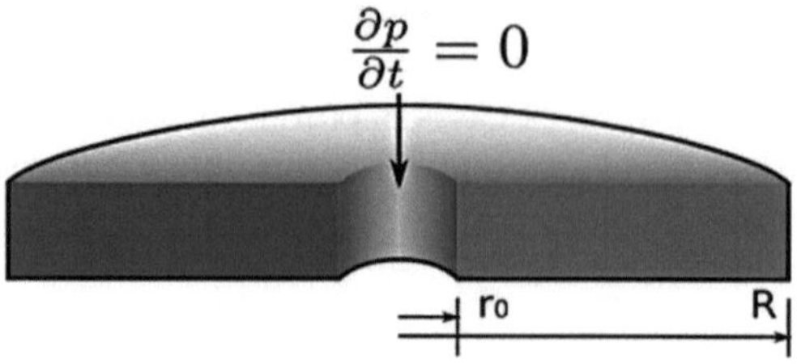

Fig. 4.19 Geometry of the homogeneous benchmark model domain

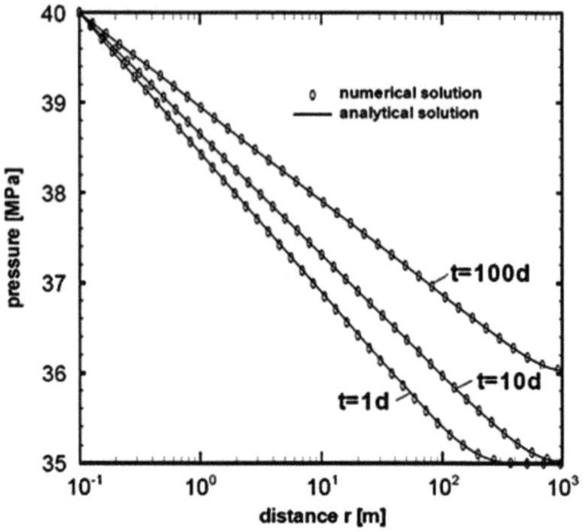

Fig. 4.20 Comparison of numerical and analytical results of the benchmark test example

the fluid properties can be derived by equations of state. The cubic Peng-Robinson EoS was applied and the viscosity correlation developed by Fenghour et al. (1998).

Figure 4.20 shows a comparison of analytical and numerical results of the described benchmark problem. The analytical solution is constrained to a linearized behaviour of fluid properties (gas compressibility is assumed to be constant). At high pressures (like in this example), this assumption seems to be valid, since the results of numerical and analytical solutions are almost identical. However, the plot shows that the governing equation Eq. 4.3 of the numerical model is valid and can be used to simulate the real gas behaviour during transport processes in the subsurface.

4.4.3 Geochemical and Reactive Transport Modelling of Fluid-Rock Interactions Induced by Injection of CO_2 in a Depleted Gas Reservoir

The modelling of fluid-rock interactions and expected mineral alterations induced by the injection of CO_2 in a depleted gas reservoir is a challenging problem for several reasons. Firstly, it has to be stressed that all considered chemical reactions, even those involving gases and minerals, occur in the water phase. This is a quite fair assumption, given the fact that aqueous reactions are on an order of magnitude quicker than gas-rock interactions; therefore it becomes crucial, in the context of CO_2 injection and storage in a depleted gas reservoir, to evaluate the amount of liquid water actually available for chemical reactions and its chemical composition prior to the injection of CO_2. Furthermore, the complexity intrinsic to geological media and to the chemical and physical processes occurring therein, each showing different characteristic spatial and temporal scales, can be only partially captured by computable mathematical and numerical models, mainly because of lack of comprehensive thermodynamic and kinetics data rigorously valid in the range of conditions (temperature, pressure, salinity of the formation fluids) of interest. This means that geochemical modelling has to be applied upon strongly simplified averaged, representative volumes and properties; but also that the results have to be evaluated carefully as well as, given the absence of possible calibrations, the uncertainty associated with predictions. The comparison of different thermodynamic databases and the integration of site-specific data and parameters from experiments on core samples are therefore utterly important as well as the continuous development of numerical codes.

4.4.3.1 Geochemical Characterisation of the Reservoir and Chemistry of Formation Waters

The chemical composition of the formation fluids was determined through analysis of several samples of formation fluids made available by GDF SUEZ or found in independently published research papers (Lüders et al. 2010). Dissolved concentrations are dominated by Cl, Na, Ca and K. The elements Mg, Mn and Br are also present in consistent amounts, as well as dissolved Fe. Total salinity of available samples ranges from 240 to 380 g L^{-1}; a representative average salinity of the formation was chosen at approximately 340 g L^{-1}, resulting in an ionic strength of around 7.5 molal. For such a solution the best suited activity model is that initially developed by Pitzer (Harvie et al. 1984), which has the drawback of lacking parameters for relevant specific interactions, mainly involving Al and Si, and not being consistently parameterised for temperatures higher than 25 °C (the reservoir has a temperature of around 120 °C). However, since the Debye-Hückel approach gives implausible results for such salinities, it was chosen to limit the geochemical model to elements contained in available Pitzer databases. An uncertainty analysis was furthermore performed in order to evaluate the confidence in the geochemical simulations resulting from the lack of reliable thermodynamical data under the conditions found in the Altmark reservoirs. Relative errors up to 10–15 % can be easily expected (De Lucia et al. 2011), which quantitatively compares with other uncertainties associated for example to the unknown spatial variability of the reservoir.

As pointed out by the detailed mineralogical and geochemical analyses carried out on core samples, the Rotliegend sandstones of the Altmark can be divided into two main lithotypes: (1) aeolian/dry sediments, showing enhanced porosity and

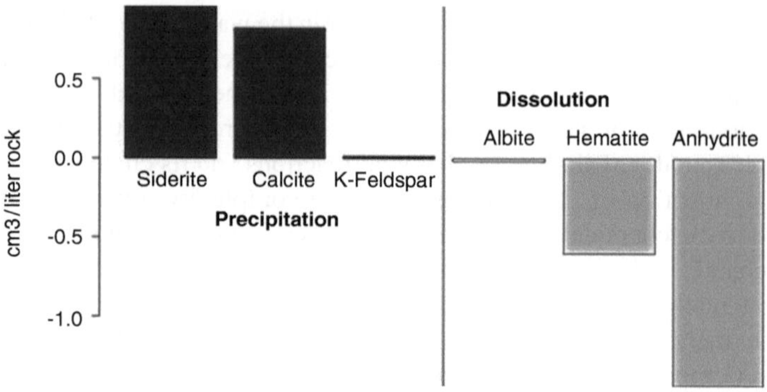

Fig. 4.21 Predicted mineral alterations due to injection of CO_2, under the hypothesis of instantaneous equilibrium. Acidizing of the solution triggers the dissolution of sulphate cement, hematite and minor amounts of albite and illite, followed by precipitation of calcite, iron carbonates and K-feldspar. The net volumetric effect is a moderate increase of porosity

permeability due to secondary bleaching/alteration and (2) aquatic/fluvial channel fill, generally showing low porosity and being less prone to bleaching. The first lithotype shows the best reservoir characteristics and is likely to be the most interesting facies for EGR operations. However, the difference between the unaltered, red sandstones and the bleached, grey ones is mainly represented by higher porosity in the latter, correlated to the partial dissolution of sulphate cementation and the presence of Fe-rich chlorites. In the simplification needed for the geochemical modelling, such average difference can be modelled by different relative abundance of mineral phases in the given representative volume. The lack of specific thermodynamical data does not allow to reliably capture in the modelling the observed complexity of clay and feldspar minerals, which are represented in the model by pure end members.

A mineral assemblage comprising K-feldspar, albite, Fe-bearing illite, hematite, and a pore-filling cementation constituted by calcite and anhydrite matches well the hypothesis of equilibrium with the composition of formation water and was taken as initial state for the geochemical modelling. The simulation for evaluation was made using the simulator PHREEQC (Parkhurst and Appelo 1999) and a Pitzer database provided by Quintessa Ltd (Benbow et al. 2008). Quite strong reducing conditions (low reduction potential) have to be assumed in the model in order to match the amount of observed dissolved Fe. Other mineral phases, which have also been observed in the core samples, like chlorite, are oversaturated, but not included in this initial equilibrium; this fact can be interpreted as the kinetic limitation of the reaction involved. The same consideration applies to K-feldspar, explaining the fact that at equilibrium the modelled dissolved K is quite lower than the measured one. Finally, a small partial pressure of gaseous CO_2 is applied, around 0.05 MPa, corresponding to the actual CO_2 fraction in the reservoir, which fixes the pH and the amount of inorganic dissolved carbon (Fig. 4.21).

Batch simulations are able to capture at least qualitatively the possible reaction path triggered by the injection of dry CO_2 in the reservoir. Given the nature of the gas reservoir, formation water is present in residual saturation, hence virtually immobile; therefore no mass transport of reactive species is expected in the liquid phase. For the following simulations, a porosity of 10 % and a high end initial water saturation of 30 % were assumed in order to scale the batch system. Simulations under the hypothesis of instantaneous equilibrium depict to some extent a worst case scenario, where all the available minerals are allowed to participate simultaneously to the reactions.

For such a system, the increase of CO_2 partial pressure to 5.0 MPa triggers the dissolution of the anhydritic cement and of hematite, followed by the precipitation of calcite and iron-carbonate (allowed to precipitate as secondary mineral). Overall volumetric effect, at equilibrium, is a moderate enhancement of porosity which can reach for this example 35–40 cm^3/L rock. This value, as already pointed out, is very high. Other predicted reactions are the dissolution of albite and on the contrary the formation of illite. In solution, the concentrations of total dissolved Fe, Al and S increase substantially, whereas K and Mg decrease. These results qualitatively agree with the first results from the batch experiments.

4.4.3.2 Development and Application of the Coupled OpenGeoSys-ChemApp Code for the Numerical Simulation of Kinetically Limited Fluid/Mineral Reactions

Dissolution and/or precipitation of minerals due to an injection of CO_2 results in a modification of the solid phase volume and hence in temporal changes of flow and transport effective porosities as well as fluid phase permeabilities. The close coupling of multiphase hydraulics, species transport and geochemical interactions between the pore fluids and mineral matrix is complicated by the very disparate time scales of individual geochemical reactions and transport processes. For an accurate numerical simulation and prediction of fluid flow and reactive transport induced by the injection of CO_2 in a depleted gas reservoir, the kinetic limitation of individual fluid-mineral interactions should be taken into account, especially, if reaction time scales significantly exceed transport time scales. The open source coupled multiphase flow and multispecies reactive transport simulator OpenGeoSys-ChemApp (OGS-CA) has been developed and extended to take into account the kinetic nature of fluid/mineral reactions and the resulting feedbacks on flow and transport from porosity changes. OGS uses the finite element method to solve the set of partial differential equations describing fluid flow and species transport. Equilibrium reactions are quantified by ChemApp using the Gibbs free energy minimisation approach (Petersen and Hack 2007a, b). The coupled OGS-CA code employs a sequential operator splitting scheme to solve the coupled processes during a simulation time step (Fig. 4.22). First, fluid flow and multispecies transport are solved by OGS. Next, the set of slow kinetic mineral reactions is quantified within OGS using a generalised formulation of a Lasaga type kinetic rate law

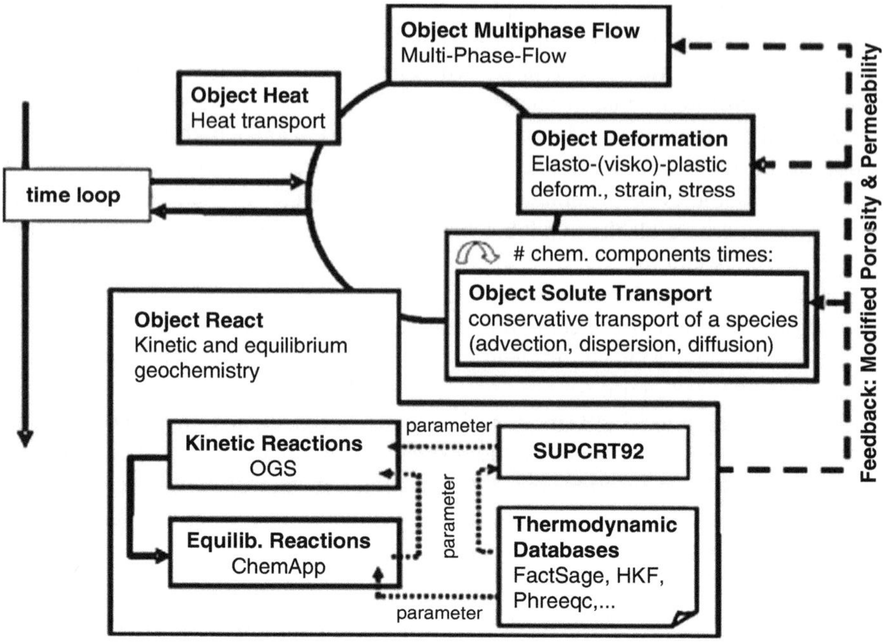

Fig. 4.22 Object oriented operator splitting concept for coupled reactive transport in OpenGeoSys-ChemApp

(Steefel and Lasaga 1994; Palandri and Kharaka 2004). Subsequently, ChemApp is used to equilibrate the geochemical system, which includes the quantification of fast (quasi-)equilibrium fluid-mineral reactions and the speciation of the solution composition. Finally, the evolution of porosities and permeabilities due to changed mineral phase volume fractions is evaluated and updated for the next time step.

The newly developed coupled simulator was carefully verified for its correctness and accuracy of the coupled reaction module by comparison against analytical solutions and other software such as PHREEQC. The verified code was applied to study short to medium-term effects (25 years) of an injection of CO_2 into the Altensalzwedel reservoir on the mineral composition and porosity evolution of a typical sandstone facies.

The overall geochemical behaviour of the system is very sensitive to the initial composition of liquid and solid phases, which, as pointed out before, are both uncertain and highly variable in space. The reactive transport model assumes a stagnant water phase at residual saturation in a low permeability region of the reservoir, which is exposed to a diffusive influx of CO_2 from the gas phase. Simulations assuming instantaneous equilibrium between fluid and all mineral phases show rapid changes in the mineral composition (dissolution of calcite and some primary silicates, precipitation of secondary carbonates) and a slow progress of distinct reaction fronts. Introducing kinetic limitations on the mineral reactions has a strong influence on the predicted behaviour of the geochemical system.

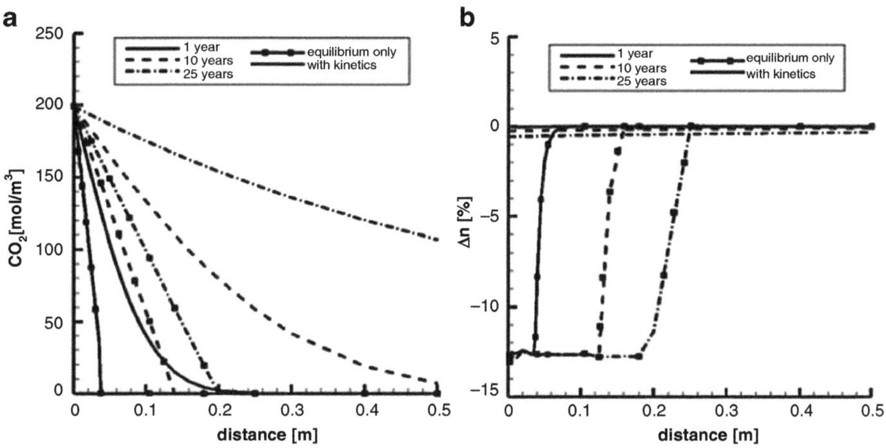

Fig. 4.23 (**a**) Comparison of simulated CO_2 dissolution fronts after 1, 10 and 25 years for a full equilibrium and kinetically limited reactive transport model; (**b**) comparison of changes in rock porosity due to mineral precipitation

Figure 4.23a shows profiles of dissolved CO_2 concentrations in the brine after 1, 10 and 25 years. Slower reaction rates allow an increased spreading and smearing of the CO_2 dissolution front. Consequently, the region in which mineral alterations are observed is significantly extended in comparison to a full equilibrium simulation. Moreover, calcite precipitation is observed instead of dissolution, as precipitation of other carbonate minerals is suppressed by the comparably slow kinetics of these reactions. In Fig. 4.23b the effect of mineral precipitation/dissolution on rock porosity is shown. For the equilibrium simulation, the progress of the mineral alteration region is clearly evident as porosity is reduced by up to about 12 % from its initial value, mainly due to carbonate precipitation. For the kinetic model, the porosity reduction by less than 1 % is only marginal, but spread over a much wider region. These strong differences found for the scenario simulations highlight the importance of accounting for realistic reaction rates in numerical simulations of CO_2-rock interactions during the injection and storage of CO_2.

4.4.3.3 Solubility Model and Dry-Out Effects

When modelling the impact of CO_2 injection on mineral alteration processes in reservoirs, as well as solubility and long-term mineral trapping or dry-out effects in the near-well region, a crucial role is played by the phenomenological model used to assess the phase equilibrium between the gas or supercritical CO_2-rich phase and the formation water, which is governed by temperature, partial pressure of the CO_2-rich phase and the actual chemical composition of the formation brine. The available mutual solubility models, as implemented in current simulators (e.g. Spycher et al. 2003), do not take into account the specific chemical composition of the brine or are not valid over a broad range of temperature and pressure, generally resulting in an overestimation of both the dissolved CO_2 and the evaporated water for very concentrated solutions.

Hence, it is of advantage to incorporate a rigorous thermodynamic formulation of the mutual solubility model CO_2-water directly into the chemical calculation, fixing the fugacity of the aqueous CO_2 by means of accurate equations of state (Duan et al. 1992) and, for calculating the evaporation of water into the CO_2-rich phase, correcting the saturation pressure of water vapour by the water activity. This was implemented into the finite element simulator OpenGeoSys, using the coupled programme PHREEQC as simulator for reactive chemistry, having access to the most up-to-date chemical and thermodynamic databases, including the Pitzer coefficients needed for high ionic strengths.

Under Altmark conditions, the relative porosity loss in the near-well region due to injection of dry CO_2 can theoretically reach as much as 8–12 % depending on the initial amount of brine in the rock volume (considering the initial water irreducible saturation range as well: 10–30 %). This approach does not consider capillary effects which could in theory replace the evaporated brine, thus locally increasing the precipitated salt amount. However, for a gas reservoir which is considered at irreducible water saturation, this seems a fair approximation. The total pore volume available to gas flow is on the contrary overall increased, if the precipitated salt does not clog the pore throats, which is actually impossible to predict numerically. Overall, this can produce an enhanced injectivity. Considering an initial porosity of 15 % and brine saturation of 20 %, a rock volume has to be flushed around 1,500 times to theoretically reach the complete evaporation of the water phase. Thus, salt deposition due to dry-out is likely to propagate for few tenths of metres – at the farthest – from injection well.

4.4.4 Generic Thermo–Hydro–Mechanical Modelling of CO_2 Injection

This part focuses on the analysis of stress changes and deformation of the storage formation including the stability and integrity of the cap rock considering three different primary stress regimes: (1) the extensional ($\sigma_{H/h} = 0.8\ \sigma_v$), (2) the isotropic ($\sigma_{H/h} = \sigma_v$), and (3) the compressive stress state ($\sigma_{H/h} = 1.5\ \sigma_v$). The coupled THM simulation was performed with the integrated code TOUGH2/EOS7C-FLAC3D (Rutqvist et al. 2002). EOS7C is a TOUGH2-EoS module which considers five mass components (water, brine, CO_2, tracer and NCG (CH_4 or N_2)) and can calculate the density of the real gas mixture, enthalpy and viscosity of the system H_2O–CO_2–CH_4 accurately (Oldenburg et al. 2004). A generic 3D plane model (20,000 m · 3,000 m · 100 m) was generated and discretized into 3,344 cells (Fig. 4.24). The whole model includes four horizontal rock layers from top to bottom, namely overburden (1,200 m), cap rock (100 m), gas reservoir (200 m) and baserock (1,500 m). Two wellbores for production and injection are 1,200 m apart in the horizontal direction and are drilled to a depth of 1,300 m, the imaginary top surface of the reservoir layer.

All rock layers are assumed to be elastoplastic materials characterised by isotropic medium properties. The reservoir has an initial porosity of 0.09 and permeability of 1.3 Darcy. All these properties are correlated to the effective mean stress using functions (4.6) and (4.7) and parameters from Rutqvist and Tsang (2002):

$$\phi = \phi_r + (\phi_0 - \phi_r) \cdot \exp(-a\sigma'_m) \tag{4.6}$$

$$k = k_0 \cdot \exp\left[c\left(\frac{\phi}{\phi_0} - 1\right)\right] \tag{4.7}$$

In order to consider the influence of possible plastic deformation on the permeability, an extra term is multiplied by the original correlation function (Gou 2011). Here, b is a constant, and assumed to be 2,300:

$$k = k_0 \cdot \exp\left[c\left(\frac{\phi}{\phi_0} - 1\right)\right] \cdot \exp\left(b \cdot \varepsilon_{vol}^{p}\right) \tag{4.8}$$

The entire simulation comprised 30 years production, 5 years CO_2-EGR and 10 years CO_2 storage. At first the model was initialised with a hydrostatic pore pressure distribution. The reservoir has an initial pore pressure of 13 MPa and a temperature of 53 °C with only a single phase gas. Then gas production was performed from both wells with a fixed bottom-hole pressure of 5 MPa. After the natural gas production, CO_2 was injected into the left well (Fig. 4.24) at a rate of 0.01 kg $s^{-1}m^{-1}$ (i.e. 315 t a^{-1} m^{-1} referred to the injection area of the model), while the well on the right was constantly kept in production of the gas. In the CO_2 storage phase, the right well was sealed and CO_2 further injected into the left well at a rate of 0.1 kg s^{-1} m^{-1} (3,150 t a^{-1} m^{-1}).

Figure 4.25 shows the CO_2 mass fraction in the gas mixture within the 5 years of CO_2-EGR. It is obvious how CO_2 moves down from the injection point due to its larger density compared to CH_4 at the same thermodynamic conditions. After 5 years of injection, CO_2 reaches the production well, and its mass fraction in the produced gas rises to 8 %. During the entire process the pore pressure in the reservoir did not change.

Figure 4.26 shows stress changes along the vertical line from the injection point during the CO_2 storage process. The total vertical stress slightly increases (<0.1 MPa), while the effective vertical stress decreases by 2 MPa due to the rise of reservoir pressure. However, the horizontal stress shows an opposite tendency. The horizontal effective stress slightly changes (0.3 MPa), while the total stress increases by 1.7 MPa. During the overall process no plastic deformation is observed, i.e., the reservoir and cap rock reacted elastically. The simulated values agree well with this analytical solution (Hou et al. 2009b):

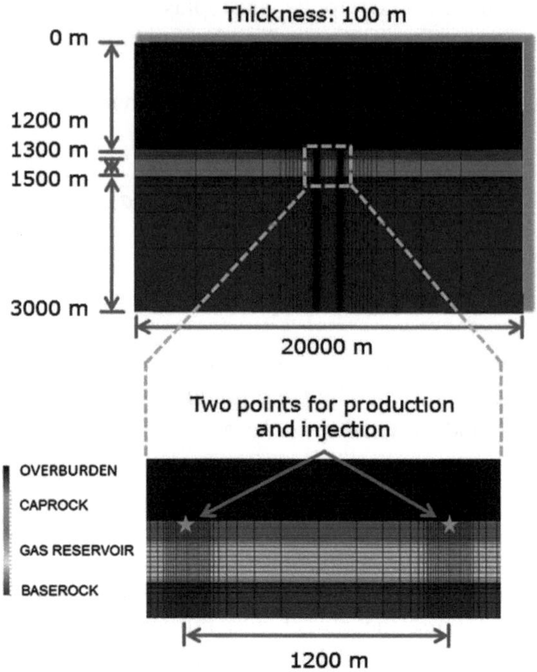

Fig. 4.24 Generic 3D plane model for the simulation

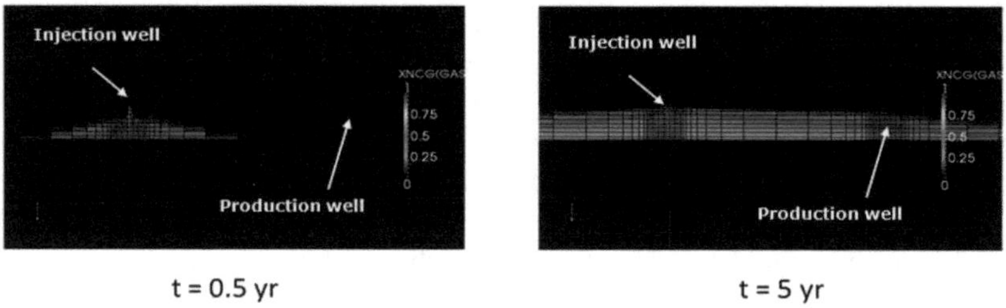

Fig. 4.25 CO_2 mass fraction in the gas mixture at different periods of time during the CO_2-EGR process

$$\Delta\sigma_{H/h} = \alpha \frac{1-2v}{1-v} \Delta p \tag{4.9}$$

Figure 4.27 shows the stress path from the injection point for three different primary stress regimes. After 45 years, none of the stress points is reached or has attained the strength of the rock with the result of fracturing (for fracture strength it is assumed that the fracture has the same friction angle as the rock but zero cohesion). The stress path of the compressive regime points toward the strength line, while the stress paths for the other two regimes evolve parallel with the strength line. With further injection, the compressive stress regime is likely to reach the fracture strength.

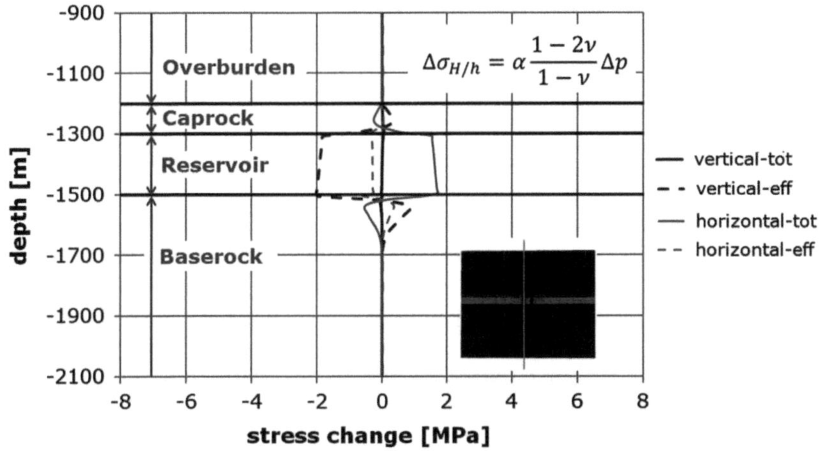

Fig. 4.26 Stress change along the vertical line through the injection point during the CO_2 storage process

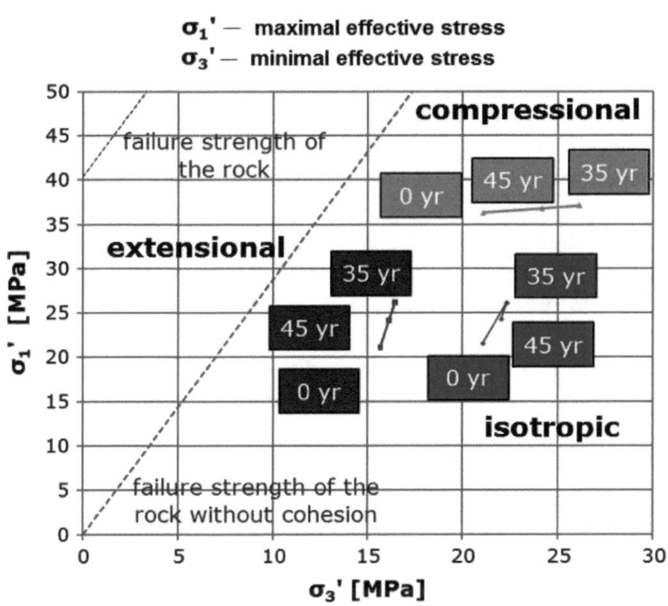

Fig. 4.27 Stress path of injection point for three different primary stress states

4.4.5 Numerical Simulation of Tracer Migration Using an Advanced Isothermal Approach for Compositional Gas Flow

Tracer tests aimed at the characterisation of geologic formations in the underground have been reported since decades (e.g. Wallick and Jenkins (1954) and literature cited therein). Within this context, detectable materials (tracer substances) are injected in a formation, and measurements of concentrations and arrival times of the tracer at observation wells are used to get information about flow paths as well as the possible presence of impermeable faults in the formation. Furthermore, analysing tracer test results, averaged values of certain physical quantities can be determined, including flow velocity, formation permeability, fluid saturation in the formation as well as diffusion and dispersivity parameters of the porous media composing the formation. Thus, tracer tests can be used equally to parameterise and to validate numerical models for the simulation of hydro-mechanical formation behaviour.

In order to obtain the favoured results from tracer tests, conservative tracer substances should be used, i.e. chemicals that do not react with other fluids or minerals, and that do not degrade. Within this context, Kr has been used successfully as tracer in the Frio-Brine-Pilot Experiment (Freifeld et al. 2005) in Liberty County, Texas (United States) and at the CO_2 storage pilot site Ketzin (Zimmer et al. 2010) in Brandenburg (Germany).

Here, a conservative inter-well tracer test for the multi-layered Altmark natural gas field is presented that has been designed to estimate the propagation velocity of CO_2 in the reservoir rocks under consideration using Kr as tracer substance. For both projects mentioned above (Frio and Ketzin), the tracer was injected in saline aquifers. In contrast, tracer injection in the Altmark is designed for gas-bearing reservoir rocks. Furthermore, significant differences of the Altmark case to the cited studies consist of increased Kr-volumes (Frio 0.024 m^3 (STP = standard conditions for temperature and pressure), Ketzin 6.25 m^3 (STP), Altmark 20 m^3 (STP)) and increased distances from injection to monitoring wells (Frio and Ketzin ca. 100 m, Altmark max. 1.5 km). Prior to the designed field test the propagation of the tracer was simulated with two different numerical models (OGS and ECLIPSE 100/300).

4.4.5.1 Design of the Tracer Field Test

The design of the field test governs some of the boundary conditions of the respective simulation. The tracer field test is designed to use 20 m^3 (STP) of Kr tracer together with 15,000 m^3 (STP) of N_2 carrier gas to be injected into the reservoir, using borehole S13 (Fig. 4.28). For operational reasons, currently only the uppermost reservoir horizons (ROF, 17B and 16B) of the gas reservoir are available for injection. Production gas from wells S16 and S14 (at a distance of 0.9 and 1.5 km from the injection well) will be sampled continuously at a rate of

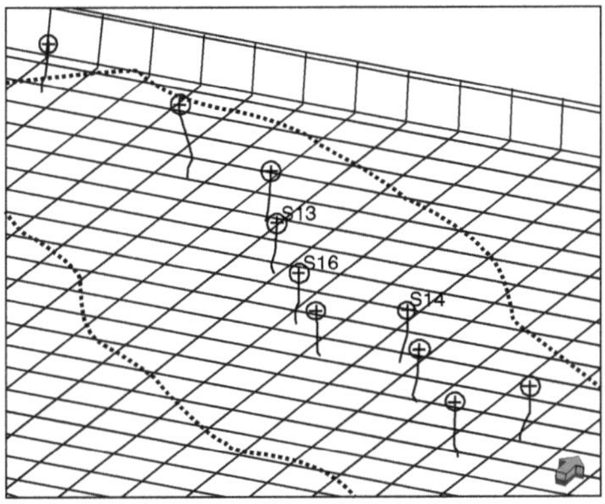

Fig. 4.28 Detail of the schematic geographic model of the Altmark natural gas field with certain boreholes, including injection well S13 and observation wells S14 and S16 for tracer migration simulation (covering an area of about 8 km^2 – 2.3 km east–west, 3.5 km north–south; according to a model provided by GDF SUEZ E&P Deutschland GmbH)

50 cm^3 h^{-1}. A portable quadrupole mass spectrometer will be used for continuous and real time gas analysis. Baseline concentrations of Kr $\leq$1.8 ppmv in the Altmark natural gas have already been determined. Thus, any significant increase of Kr concentrations will be detectable in the ppmv-range.

The virtual tracer test is simulated using a compositional gas flow model assuming isothermal conditions, i.e. only the pressure field affects the tracer transport combined with the velocity field (calculated using modified Darcy's law). The Peng-Robinson's equation of state (Peng and Robinson 1974) was used to calculate the density of the mixture (carrier gas and tracer) at given pressure and temperature (using constant values). For this calculation, critical constants (p_c, T_c) are required and the so-called acentric factor (a dipole moment as gases are polar in nature; the acentric factor is zero for Kr) of each component of the mixture. Other fluid parameters are pressure and composition dependent, for example dynamic viscosity and the hydrodynamic dispersion tensor. In order to characterise the hydrodynamic dispersion tensor, exponent, self-diffusion coefficient, tortuosity and dispersivity values are required (values applied are published by Vargaftik 1975). For mixing of the components a combination rule which needs energy and distance parameters was adopted. Details of the mathematical and numerical aspects of the used compositional gas flow model are presented in Singh et al. (2011, 2012b).

The tracer transport was simulated considering the injection of Kr as a conservative tracer into the top three reservoir layers of the Altmark natural gas field at injection well S13. The tracer migration and the arrival time at the observation wells S14 and S16 were investigated as well. The assumed Kr concentration does not change the material properties of N_2 as carrier gas significantly (e.g. density, dynamic viscosity). Specific transport processes characterising the migration of a conservative tracer are advection and diffusion-dispersion. Advection is the movement of tracer particles due to pore fluid flow caused by hydraulic gradients. The mean pore fluid velocity (calculated from Darcy's law) is used to predict advection. This is a reasonable approximation of slow laminar flow arising in reservoir engineering problems (Görke et al. 2011).

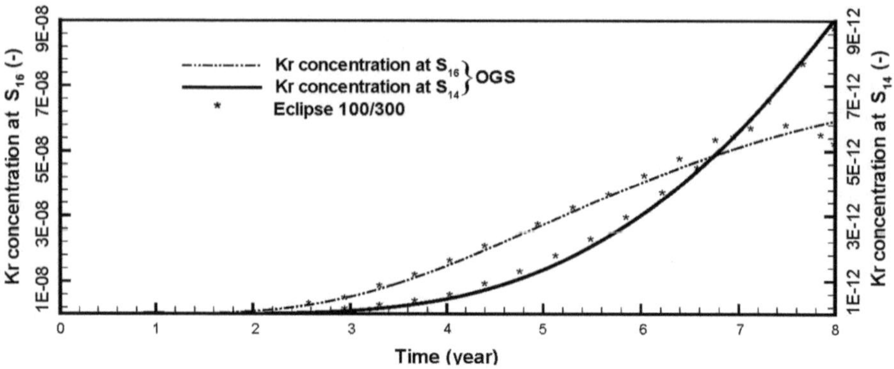

Fig. 4.29 Comparison of OpenGeoSys results for arrival of the Kr tracer at wells S14 and S16 with data gained from ECLIPSE (ECLIPSE simulation results provided by GDF SUEZ E&P Deutschland GmbH)

The designed tracer test was simulated using the compositional gas flow model implemented in the scientific open-source finite element simulator OGS (Singh et al. 2012a), whereas results have been compared with data of corresponding simulations performed by GDF SUEZ using the commercial simulator ECLIPSE. Considering the upper three reservoir layers of the formation of interest together with two intermediate cap rock layers, the problem was discretized spatially in terms of a three-dimensional domain with 12,610 hexahedral type elements. The modelling domain had a thickness of 62 m along the injection well S13, and the horizontal distances between the injector and the observation wells, taken from the geological model, were 916 m (S16) and 1,506 m (S14) respectively. Structure, geometry and parameterisation of the simulation model were defined according to the static model provided by GDF SUEZ via the CLEAN Data Management System. The history match of this model was performed using the ECLIPSE simulator.

Corresponding to real site conditions, the following initial and boundary conditions were defined for the simulation model:

- Initial conditions: The considered reservoir-cap rock system is at 4 MPa pressure before simulation starts assuming isothermal conditions (393.15 K). Pressure gradients measured by probes between wells S13, S14 and S16 were neglected. The pores of the solid skeleton are filled with 26 % CH_4 and 74 % N_2.
- Boundary conditions: No flow conditions at the outer boundary of the model domain were defined, producing natural gas from wells S14 and S16 with rates of 45,000 and 20,000 m^3 per day, respectively.
- Source term: The simulation of the virtual test was started with the injection of 15,000 m^3 (STP) of N_2 within 1 day, followed by 30 days of no injection. Afterward, 20 m^3 (STP) of Kr were injected within 1 day and a pause of another day. Finally 15,000 m^3 (STP) of N_2 were injected within 1 day accelerating transport of the tracer.

Porosity and permeability of the reservoir layers under consideration vary between n = 0.051 and 0.088 and k = 1.39 and 2.33 mD, respectively. For both cap rock layers, values of n = 0.014 and k = 0.1 mD were assumed. Density of the reservoir rock mass was assumed to be homogeneous with 2,460 kg m^{-3}, whereas for the cap rock it was varied between 2,700 and 2,940 kg m^{-3}.

In Fig. 4.29 the evolution of the tracer concentration at observation wells S14 and S16 is presented simulated for a period of about 8 years. As can be expected, the arrival time for nearer well S16 is shorter than for well S14. In addition, the tracer concentration level at well S16 is higher than that at well S14. The results of the simulators OGS and ECLIPSE are similar. At well S14, both data sets are in close agreement while at well S16, the OGS results show a slightly different behaviour after 7 years (Singh et al. 2012b).

References

Bell JS, Gough DI (1979) Northeast-southwest compressive stress in Alberta: evidence from oil wells. Earth Planet Sci Lett 45:475–482

Benbow S, Metcalfe R, Wilson J (2008) Pitzer databases for use in thermodynamic modelling. Quintessa Technical Memorandum (unpublished)

Blümling P (1986) In-Situ-Spannungsmessungen in Tiefbohrungen mit Hilfe von Bohrlochrandausbrüchen und die Spannungsverteilung in der Kruste Mitteleuropas und Australiens. Dissertation, Universität Karlsruhe, 135 p

Boles JR, Johnson KS (1984) Influence of mica surfaces on pore-water pH. Chem Geol 43:303–317

Bowker KA, Shuler PJ (1991) Carbon dioxide injection and resultant alteration of the Weber Sandstone, Rangely Field, Colorado. Am Assoc Petr Geol Bull 75:1489–1499

Braun E (2009) Determination with RACOS® of the 3D in situ loadings, the pore pressure effectivenesses and the deformation parameters for several strata in the Salzwedel reservoir area. Report submitted to TU Clausthal, 64 pp

Cacace M, Bayer U, Marotta AM, Lempp C (2008) Driving mechanisms for basin formation and evolution. In: Littke R, Bayer D, Gajewski D, Nelskamp S (eds) Dynamics of complex intracontinental basins. The Central European Basin System. Springer, Berlin/Heidelberg, pp 37–66

Class H, Ebigbo A, Helmig R, Dahle HK, Nordbotten JM, Celia MA et al (2009) A benchmark study on problems related to CO_2 storage in geologic formations: summary and discussion of the results. Comput Geosci 13(4):409–434

De Lucia M, Stadler S, Kühn M (2011) Evaluating uncertainties in geochemical modeling for CO_2 storage: multivariate sensitivity analysis of database entries. Geophys Res Abstr 13: EGU2011–EGU12020

Duan Z, Møller N, Weare J (1992) An equation of state for the CH_4-CO_2-H_2O system: I. Pure systems from 0 to 1000°C and 0 to 8000 bar. Geochim Cosmochim Acta 56:2605–2617

Duan Z, Sun R, Zhu C, Chou IM (2006) An improved model for the calculation of CO_2 solubility in aqueous solutions containing Na^+, K^+, Ca^{2+}, Mg^{2+}, Cl, and SO_4^{2-}. Mar Chem 98:131–139

Fenghour A, Wakeham WA, Vesovic V (1998) The viscosity of carbon dioxide. J Phys Chem Ref Data 27:31–44

Fischer C, Dunkl I, Wijbrans JR, von Eynatten H, Gaupp R (submitted) Thermochronological constraints on mesodiagenetic processes in Rotliegend sandstones from the Central European Basin. Geol Mag

Freifeld BM, Trautz RC, Kharaka YK, Phelps TJ, Myer LR, Hovorka SD, Collins DJ (2005) The U-tube: a novel system for acquiring borehole fluid samples from a deep geologic CO_2 sequestration experiment. J Geophys Res 110:B10203

Gaupp R (1996) Diagenesis types and their application in diagenesis mapping. Zbl Geol Paläont 1:1183–1199

Görke U-J, Park C-H, Wang W, Singh AK, Kolditz O (2011) Numerical simulation of multiphase hydromechanical processes induced by CO_2 injection in deep saline aquifers. Oil Gas Sci Technol – Revue d'IFP Energies nouvelles 66(1):105–118

Gou Y (2011) Untersuchungen zu THM-gekoppelten Reaktionen der Speicherformationen und des Deckgebirges bei CO_2-EGR mit dem integrierten Code TOUGH2 –FLAC3D. Masterarbeit am Institut für Erdöl-/Erdgastechnik der TU Clausthal

Haefner F, Sames D, Voigt H-D (1992) Wärme- und Stofftransport. Mathematische Methoden. Springer, Berlin/New York

Hartmann B (1997) Mobilität von Seltenen-Erd-Elementen (SEE) und deren Fixierung in Karbonatphasen am Beispiel von Rotliegend-Sandsteinen des Norddeutschen Beckens. Ph.D. thesis, University Mainz, p 116

Harvie CE, Møller N, Weare JH (1984) The prediction of mineral solubilities in natural waters: The Na-K-Mg-Ca-H-Cl-SO_4-OH-HCO_3-CO_3-CO_2-H_2O system to high ionic strengths at 25 °C. Geochim Cosmochim Acta 48(4):723–751

Haszeldine RS, Quinn O, England G, Wilkinson M, Shipton ZK, Evans JP, Heath J, Crossey L, Ballentine CJ, Graham CM (2005) Natural geochemical analogues for carbon dioxide storage in deep geological porous reservoirs, a United Kingdom perspective. Oil Gas Sci Technol – Rev IFP 60:33–49

Hillier S (1994) Pore-lining chlorites in siliciclastic reservoir sandstones: electron microscope, SEM, and XRD data, and implications for their origin. Clay Miner 41:240–259

Hou MZ, Tadongmo FA, Pusch G (2009a) Secondary in-situ stresses resulting from change of pore pressure influence on the maximal storage pressure of CO_2 or natural gas in reservoirs. DGMK/ÖGEW-Frühjahrstagung, Celle

Hou MZ, Yoon J, Khan IU, Wundram L (2009b) Determination of the 3D primary stresses and hydro-mechanical properties of several strata in the Salzwedel reservoir area & multi-layer caprock-reservoir 2D-model for the Altensalzwedel test field for THMC coupled numerical modelling. Report to the BMBF project 03G0704Q, Milestones I & II of the subproject PI.2 in the TV-III of the CLEAN-Project, 25 pp

Katz DL, Legatski MW, Tek MR, Gorring L, Nielson RL (1966) How water displaces gas from porous media. Oil Gas J 64:55

Kirnbauer T (1997) Die Mineralisation der Wiesbadener Thermalquellen. Jahrb Nass Ver Naturk 118:5–90

Kirste DM, Watson MN, Tingate PR (2004) Geochemical modeling of CO_2-water-rock interaction in the Pretty Hill Formation, Otway Basin. In: Boult PJ, Johns DR, Lang SC (eds) Eastern Australasian Basins symposium II, Adelaide/South Australia, 19–22 Sept. 2004. The Petroleum Exploration Society of Australia (PESA), Special publication, pp 403–411

Lüders V, Plessen B, Romer RL, Weise SM, Banks DA, Hoth P, Dulski P, Schettler G (2010) Chemistry and isotopic composition of Rotliegend and Upper Carboniferous formation waters from the North German Basin. Chem Geol 276:198–208

Lutz M, Kaasschieter JPH, van Wijhe DH (1975) PD 2(3) geological factors controlling Rotliegend gas accumulations in the Mid-European Basin. 9th World Petroleum Congress, 11–16 May 1975, Tokyo, Paper WPC 16109, pp 93–103

May F (2005) Alteration of wall rocks by CO_2-rich water ascending in fault zones: natural analogues for reactions induced by CO_2 migrating along faults in siliciclastic reservoir and cap rocks. Oil Gas Sci Technol – Rev IFP 60:19–32

Moeck I, Schandelmeier H, Holl G (2009) The stress regime in a Rotliegend reservoir of the Northeast German Basin. Int J Earth Sci (Geol Rundsch) 98:1643–1654

Moog H, Mönig H (2009) Erstellung von Parameterdateien für die Verwendung mit CHEMAPP und PHREEQC. Final report to BGR project 45–4500046190, 79 pp

Norden B, Förster A, Behrends K, Krause K, Stecken L, Meyer R (2012) Geological 3D model of the larger Altensalzwedel area, Germany, for temperature prognosis and reservoir simulation. Environ Geo. sci 67(2):511–526 doi:10.1007/s12665-012-1709-9

Oldenburg CM, Moridis GJ, Spycher N, Pruess K (2004) EOS7C Version 1.0: TOUGH2 module for carbon dioxide or nitrogen in natural gas (methan) reservoirs. Lawrence Berkeley National Laboratory, LBNL-56589

Palandri JL, Kharaka YK (2004) A compilation of rate parameters of water-mineral interaction kinetics for application to geochemical modeling: U.S. Geological Survey Water-Resources Investigations report 04–1068

Parkhurst DL, Appelo CAJ (1999) User's guide to PhreeqC (Version 2) – a computer program for speciation, batch-reaction, one-dimensional transport, and inverse geochemical calculations. U.S. Geological Survey, Water-Resources Investigations report 99–4259, Denver

Peng DY, Robinson DB (1974) A new two-constant equation of state. Ind Eng Chem Fundam 15:59–64

Pérez-Garrido C, Astilleros JM, Fernández-Díaz L, Prieto M (2009) In situ AFM study of the interaction between calcite 1014 surfaces and supersaturated Mn^{2+}-CO_3^{2-} aqueous solutions. J Crystal Growth 311:4730–4739

Petersen S, Hack K (2007a) The thermochemistry library ChemApp and its applications. Int J Mater Res 98:268–277

Petersen S, Hack K (2007b) The thermochemistry library ChemApp and its applications. Int J Mater Res 98(10):935–945

Pitzer KS, Peiper JC, Busey RH (1984) Thermodynamic properties of aqueous sodium chloride solutions. J Phys Chem Ref Data 13:1–102

Pudlo D, Gaupp R (2010) Mineralalterationsphänomene im mittleren und oberen Buntsandstein des Thüringer Beckens. In: Annual meeting of the Thuringian Geological Association, Jena, 5–6 June 2010. Conference proceedings, p 47

Reitenbach V, Pusch G, Möller M, Koll S, Costantini A, Junker A, Anton H (2007) Ein rechnergestützter Gesteinsdatenkatalog für Gasformationen. DGMK/ÖGEW-Frühjahrstagung, Celle, 2007, 11 pp

Röckel T, Lempp C (2003) Der Spannungszustand im Norddeutschen Becken. Erdöl Erdgas Kohle 119(2):73–80, Urban Verlag Hamburg/Wien

Ross G (1982) The dissolution effects of CO_2-brine systems on permeability of UK and North Sea calcareous sandstones. Paper SPE/DOE 10685 presented at third joint symposium on enhanced oil recovery of the society of petroleum engineers, Tulsa, 4–7 Apr 1982, pp 154–162

Rutqvist J, Tsang C-F (2002) A study of caprock hydromechanical changes associated with CO_2 injection into a brine aquifer. Environ Geol 42:296–305

Rutqvist J, Wu Y-S, Tsang C-F, Bodvarsson G (2002) A modeling approach for analysis of coupled multiphase fluid flow, heat transfer, and deformation in fractured porous rock. Int J Rock Mech Min Sci 39:429–442

Sayegh SG, Krause FF, Girard M, DeBree C (1990) Rock/fluid interactions of carbonated brines in a sandstone reservoir: Pembina Cardium, Alberta, Canada. SPE Formation Eval 5:399–405

Shiraki R, Dunn TL (2000) Experimental study on water-rock interactions during CO_2 flooding in the tensleep formation, Wyoming, USA. Appl Geochem 15:265–279

Singh AK, Görke U-J, Kolditz O (2011) Numerical simulation of non-isothermal compositional gas flow: application to carbon dioxide injection into gas reservoirs. Energy 36:3446–3458

Singh AK, Baumann G, Henninges J, Görke U-J, Kolditz O (2012a) Numerical analysis of thermal effects during carbon dioxide injection with enhanced gas recovery: a theoretical case study for the Altmark gas field. Environ Earth Sci. 67(2):497–509 doi:10.1007/s12665-012-1689-9

Singh AK, Pilz P, Zimmer M, Kalbacher T, Görke U-J, Kolditz O (2012b) Numerical simulation of tracer transport in the Altmark gas field. 67(2):537–548 Environ Earth Sci. doi:10.1007/s12665-012-1688-x

Span R, Wagner W (1996) A new equation of state for carbon dioxide covering the fluid region from the triple-point temperature to 1100 K at pressures up to 800 MPa. J Phys Chem Ref Data 25(6):1509–1596

Spycher N, Pruess K, Ennis-King J (2003) CO_2-H_2O mixtures in the geological sequestration of CO_2. I. Assessment and calculation of mutual solubilities from 12 to 100°C and up to 600 bar. Geochim Cosmochim Acta 67(16):3015–3031

Steefel CI, Lasaga AC (1994) A coupled model for transport of multiple chemical species and kinetic precipitation/dissolution reactions with application to reactive flow in single phase hydrothermal systems. Am J Sci 294:529–592

Vargaftik NB (1975) Tables on the thermophysical properties of liquids and gases, 2nd edn. Hemisphere, New York

Wallick GC, Jenkins R (1954) Analysis of short-time tracer injection in underground formations. J Appl Phys 25(12):1491–1496

Wang Z (2000) Dynamic versus static elastic properties of reservoir rocks. In: Wang Y, Nur A (eds) Seismic and acoustic velocities in reservoir rocks, vol 3, Recent development. Society of Exploration Geophysicists, Tulsa, p 9

Zimmer M, Erzinger J, Kujawa C (2010) CO_2-SINK Group. The gas membrane sensor (GMS): a new method for gas measurements in deep boreholes applied at the CO_2-SINK site. Int J Greenhouse Gas Control 5(4):995–1001

Chapter 5
Reservoir and Cap Rock Monitoring

Michael Kühn, Johannes AC Barth, Gunther Baumann, Veith Becker, Michaela Bock, Stefan Buske, Ralf Fritschen, Rüdiger Giese, Christine Groß, Jan Henninges, Lars Houpt, Dagmar Kock, Martin Krüger, Daria Morozova, Anssi Myrttinen and Hilke Würdemann

Abstract One aim of the CLEAN project was to develop and test monitoring methods for the reservoir cap rock and the reservoir itself. It is shown here that advanced injection and production profile evaluation can be achieved using a combination of pressure, temperature and spinner flow meter data. Using distributed temperature sensing, temperature profiles in gas-filled wells can be acquired, as the sensor cable can be stationary during the measurement allowing for simultaneous thermal equilibration along the entire logged profile. A first field test of the developed hybrid wireline logging system was successfully performed under static conditions and the feasibility of warm-back monitoring was shown based on the results of numerical simulations for a possible CO_2 injection scenario. A combined approach using the developed hybrid system for enhanced production

M. Kühn (✉) • G. Baumann • M. Bock • R. Giese • J. Henninges • D. Morozova • H. Würdemann
Helmholtz Centre Potsdam, GFZ - German Research Centre for Geosciences, Telegrafenberg, 14473 Potsdam, Germany
e-mail: michael.kuehn@gfz-potsdam.de

J.A.C. Barth • V. Becker • A. Myrttinen
GeoZentrum Nordbayern Friedrich-Alexander-University, Erlangen-Nürnberg, Schlossgarten 5, 91054 Erlangen, Germany

S. Buske
TU Bergakademie Freiberg, 09588 Freiberg, Germany

R. Fritschen
DMT GmbH & Co KG, 45307 Essen, Germany

C. Groß
DMT GmbH & Co KG, 45307 Essen, Germany

Freie Universität Berlin, Fachrichtung Geophysik, Malteserstrasse 74-100, 12249 Berlin, Germany

L. Houpt
Freie Universität Berlin, Fachrichtung Geophysik, Malteserstrasse 74-100, 12249 Berlin, Germany

D. Kock • M. Krüger
Federal Institute for Geosciences and Natural Resources (BGR), 30655 Hannover, Germany

M. Kühn and U. Münch (eds.), *CLEAN: CO_2 Large-Scale Enhanced Gas Recovery in the Altmark Natural Gas Field*, Advanced Technologies in Earth Sciences,
DOI 10.1007/978-3-642-31677-7_5, © Springer-Verlag Berlin Heidelberg 2013

logging during injection followed by warm-back monitoring within a subsequent shut-in period would allow for accurate determination of the spatial extent and injectivity of individual CO_2 injection intervals.

Application of the PNG (pulsed neutron gamma) method for estimation of saturation changes is hampered under the considered conditions because of the low contrast between CO_2 and natural gas. However, considering the effect of drying-out of the bound water with associated salt precipitation, which is expected to occur close to injection wells, could be monitored with PNG logs depending on the volume of bound brine.

In preparation to seismic MSP/VSP experiments numerical models were calculated, which helped to find boundary conditions for the survey design. In this framework it was possible to identify the receiver depths with largest amplitude changes (more than 5 %) depending on the offset, the dimension of density and velocity changes and the expansion radius of the CO_2 front. As expected, the amplitude changes during the replacement of the reservoir gas by CO_2 are very small. For a direct detection of the CO_2 front using the seismic wave field, large pressure and/or temperature changes would be required. The concentration of an active seismic monitoring in the Altmark must be focused on the appearance of possible leakages, which cause larger velocity changes. Tests with synthetic datasets showed that seismicity location with the diffraction summation method is possible. For this purpose four stations or rather four boreholes are necessary.

For CO_2 injection projects, stable isotopes and isotope mass balance calculations are new and promising tools to monitor migration of CO_2 plumes and to accurately quantify the amount of CO_2 dissolved in water respectively. The baseline monitoring of the Altmark site yielded a good overview of baseline variations of the DIC (dissolved inorganic carbon) and isotope values, which are clearly distinguishable from the CO_2 to be injected. Laboratory results revealed theoretical trends. They indicate that with high pCO_2, $\delta^{13}C_{DIC}$ and $\delta^{18}O_{H_2O}$ add up as valuable monitoring tools for different respective intervals of DIC concentrations. In these applications of isotope monitoring, fractionation plays a major role as they influence the isotope ratios resulting from dissolution and speciation processes. Thus, the determination of their temperature and pressure dependence as well as the influence of salinity are important to hone isotope techniques to serve as accurate tools for the assessment of CCS reservoirs before and after injection.

The microbial community of the 3.5 km deep nearly depleted gas reservoir was analysed by molecular genetic techniques. Comparative analysis of the fluid samples from three different wells, which differ in their temperature profiles and operative history are presented. DNA fingerprinting indicated the presence of microorganisms similar to previously identified microbes from thermophilic and anaerobic environments in deep hypersaline and hot reservoir environments. Phylogenetic analyses revealed a high microbial diversity including different H_2-oxidising bacteria (hydrogenophaga, acidovorax, ralstonia and pseudomonas), thiosulfate-oxidising bacteria (diaphorobacter), dissimilatory metal reducers (pantoea), aromatic-degrading deep sediment inhabitants, (sphingomonas), extremophiles (bacillus), and biocorrosive thermophilic microorganisms. Additionally several sequences of microorganisms similar to representatives from different saline, hot, anoxic, deep environments were detected, which had not been cultivated previously. Cell numbers of SYBR Green total cell counts were less than 102 cells/ml and the cells were usually attached to particles.

5.1 Well Logging for Injection and Saturation Profiling

During injection of CO_2 in gas fields, both for enhanced gas recovery or geological storage of CO_2, monitoring of the fluid movement induced in the subsurface is required. Classic production logging (PL) includes downhole measurement of pressure, temperature, and fluid velocities, e.g. with a spinner flow meter, to estimate flow rates and phase composition within a producing or injecting well. The PL data is used to determine the location of injection or production intervals within the reservoir and the individual flow rates and quality of the fluids. Specifically under conditions representative for CO_2 injection at the Altmark site, monitoring of the thermal disturbance caused by fluid injection (warm back) could also be used to identify injection intervals. But in gas-filled wells wireline temperature logs over longer distances are virtually impossible, because the thermal equilibration time of conventional temperature sensors in gas is slow, requiring very low logging speeds and therefore long measurement times. Within this study a new hybrid wireline logging system, which enables simultaneous deployment of classic PL tools and acquisition of accurate and reliable temperature data in gas wells, was designed and tested during a field experiment.

Another possible option to observe CO_2 injection at the borehole scale is to monitor saturation changes using pulsed neutron-gamma (PNG) logging, which is widely used for saturation evaluation in oil and gas fields. The PNG method has already been applied successfully for monitoring of CO_2 injection in saline aquifers (e.g. Sakurai et al. 2005; Vu-Hoang et al. 2009). But since the contrast of physical properties between gases is low, a decreased sensitivity to evaluate saturation changes in depleted gas fields can be expected. Therefore, in order to assess the potential of the PNG method for conditions similar to the Altmark site, a sensitivity analysis was performed within this study.

5.1.1 Temperature Monitoring

The temperature log is classic method for determination of formation temperatures, e.g. for estimation of heat flow via the geothermal gradient and thermal conductivity. It can also be used to locate fluid movement, or in order to identify hydraulically fractured zones along a well. Here it is most effective, if changes over time can be monitored with short repeat times, because changes by advection are usually rapid, as opposed to changes by conduction which are much slower. Traditionally, temperature profiles are measured using wireline logging, with a temperature sensor inside a logging tool, which is lowered into the well across the zone of interest. The method of fibre-optic distributed temperature sensing (DTS) enables registration of temperature profiles over distances of 20 km and longer with high spatial and temporal resolution. One of the main advantages here evolves during monitoring of dynamic processes involving rapid temperature changes: With wireline logging, several logging runs are required. Depending on the length of the interval which has to be covered, each individual logging run may require up to several hours, and the individual data points of the profiles are not recorded at the same time. Using DTS, profiles can be registered almost instantaneously over long distances and changes over time can be conveniently monitored once the sensor cable is in place.

In the past 10 years there is a rapidly growing number of DTS applications in the petroleum industry, e.g. with permanent downhole sensors for production monitoring in wells which are not accessible with wireline or coiled tubing, e.g. (Brown et al. 2000). Within an earlier study, a prototype slickline DTS sensor cable has been used to assess the effect of water frac stimulation treatments (Henninges et al. 2005). Here we present the design and first field test of a hybrid wireline logging system, which combines both classic electric wireline logging tools for PL and depth correlation, as well as optical fibres included in the logging cable for acquisition of DTS temperature profiles.

5.1.1.1 The Hybrid Wireline Logging System

The new 5,500 m hybrid wireline logging cable contains both electrical conductors and steel tubes for inclusion of the optical fibres. Multi-mode fibres are contained for DTS measurements, and single-mode fibres are included in order to allow for optional other fibre-optic methods and sensors to be used. The electrical tools include pressure, temperature and a spinner flow meter, as well as gamma ray (GR) and casing collar locator (CCL) for depth correlation. A custom-designed cable head accommodates the splice for the bottom turnaround to allow for a loop configuration with both ends of the optical fibre connected to the surface. This configuration is favourable for DTS monitoring and results in a total fibre length of 11,000 m. For the optical fibres, a polyimide/carbon coating was selected to allow for the required stability and resistance to chemical degradation (e.g. hydrogen ingression) of the fibres at elevated temperatures. The maximum operating temperature of the current system is limited by the electronic tools, which can tolerate up to 150 °C.

For selection of the cable armour, special attention was given to corrosion of steel in highly saline brines at high temperatures. Standard galvanized steel wire was selected, because it is less susceptible to stress corrosion cracking than alloys with higher chromium or nickel content. A higher degree of weight-loss corrosion, where the metal surface is more or less uniformly attacked and degradation of the mechanical stability is more predictable, has nevertheless to be accepted here.

5.1.1.2 Field Test

As the planned field measurements at the Altmark site could not be performed due to the delay of CO_2 injection, the field test was performed at the in-situ Geothermal

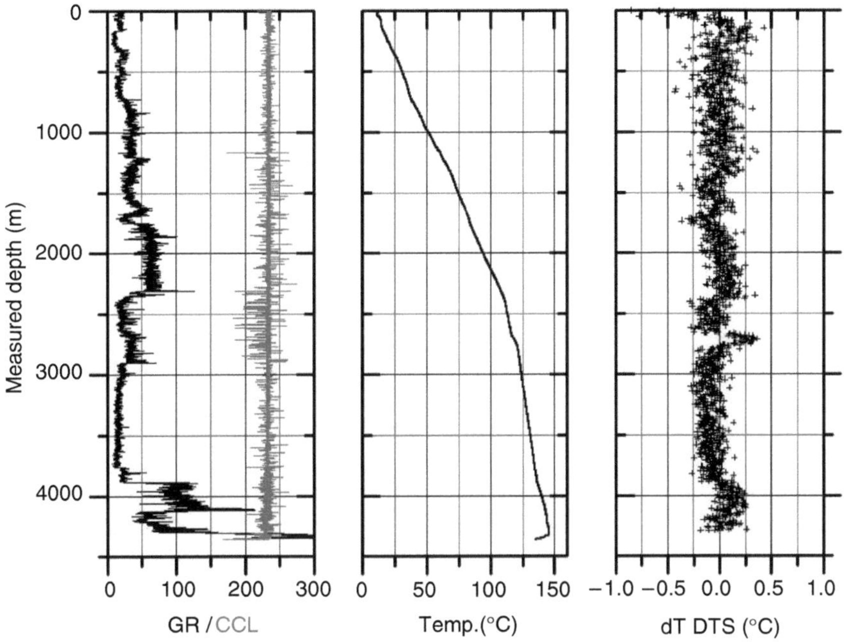

Fig. 5.1 Log data measured during the field test in the Gt GrSk 4/05 well of the geothermal test site Groß Schönebeck. *Left panel*: gamma ray (GR, *black*) and casing collar locator (CCL, *grey*). *Middle panel*: wireline log temperature profile (Temp.). *Right panel*: temperature differences between calibrated DTS and wireline temperature profiles (dT DTS)

Lab Groß Schönebeck close to Berlin, Germany. The well Gt GrSk 4/05 enabled testing of the equipment under conditions similar to the Altmark site, with a well depth of 4,400 m, maximum temperatures of up to 142 °C, and highly saline formation fluids with a content of total dissolved solids of up to 265 g L^{-1}. Nevertheless, in contrast to the Altmark site, only minor amounts of gas occur at Groß Schönebeck, and the wells are filled with brine. In November 2010 the first baseline logging runs with the hybrid logging system were performed under static conditions. The wireline logs were acquired measuring downward with a logging speed of 10 m min^{-1}. Subsequently the logging cable remained stationary at 4,300 m depth for acquisition of the DTS data.

The raw DTS data was averaged over 30 min acquisition time and calibrated using the wireline temperature data. The differences between the calibrated DTS and the wireline temperature data generally scatter around zero with a standard deviation of 0.13 °C (Fig. 5.1). Individual zones with higher offsets between the DTS and wireline data may be caused by systematic differences between the individual measurement methods, e.g. the equilibration time of the wireline sensor. The temperature resolution of the DTS data depends on the integration time over which the data is averaged. For an integration time of 30 minutes a temperature

Table 5.1 Σ values of different fluids and the associated saturations for the baseline (S_{base}) and the two assumed scenarios (S_{sc1}, S_{sc2}). The gas saturations refer to φ_{eff}, and brine/NaCl saturations to φ_{bound}

Fluid	Σ [c.u.]	S_{base}[a] [−]	S_{sc1} [−]	S_{sc2} [−]	φ
N_2 (125 °C, 4 MPa)	2.72	0.75	0	0	φ_{eff}
CH_4 (125 °C, 4 MPa)	0.98	0.25	0	0	
CO_2 (125 °C, 4 MPa)	0.003	0	1	1	
Brine[a] (275 kppm)	187.0	1	1	0	φ_{bound}
NaCl (275 kppm)	164.8	0	0	1	

[a]Data according to GDF SUEZ (2009)

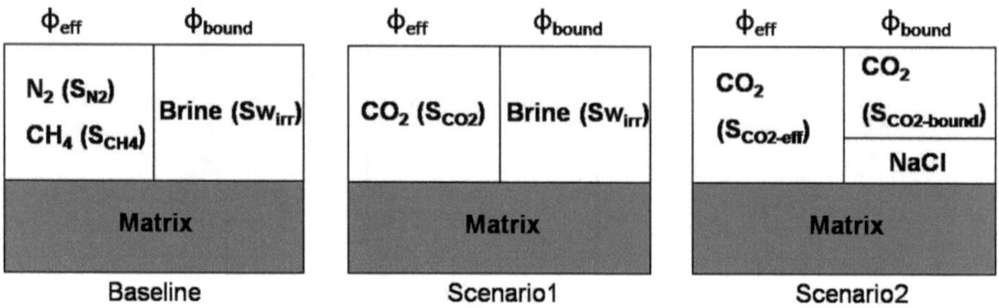

Fig. 5.2 Saturation models. Baseline: in-situ saturation condition before injection. Scenario 1: physical gas displacement in φ_{eff}. Scenario 2: physical gas displacement in φ_{eff} and complete drying out of bound water volume φ_{bound} associated with salt precipitation

resolution of up to 0.06 °C was determined, which shows that even small temperature differences can be resolved along this rather large profile of 5,500 m length.

5.1.2 *Pulsed Neutron-Gamma Logging*

During pulsed neutron-gamma logging (PNG), the macroscopic capture cross section (Σ) of the formation is measured (e.g. Smolen 1996). It is determined as the volume-weighted sum of the individual components of the matrix and the fluids contained in the pore space. With regard to the pore fluids, the capture process is mainly influenced by the presence of hydrogen and chlorine, which give rise to the high Σ values of brine (Table 5.1).

Two different mechanisms that lead to saturation changes can be distinguished (Fig. 5.2). On the one hand, physical displacement of the in-situ fluid within the pore space open for advective flow (i.e. the effective porosity) is occurring (Scenario 1). On the other hand, injection of dry CO_2 leads to dehydration of the bound formation brine, which is associated with the precipitation of salts dissolved in the brine (Scenario 2).

The irreducible water saturation (i.e. the volume of bound water) was determined from the available logging data (GDF SUEZ 2009, Fig. 5.3) and ranges from 0 % to 40 % of the total porosity.

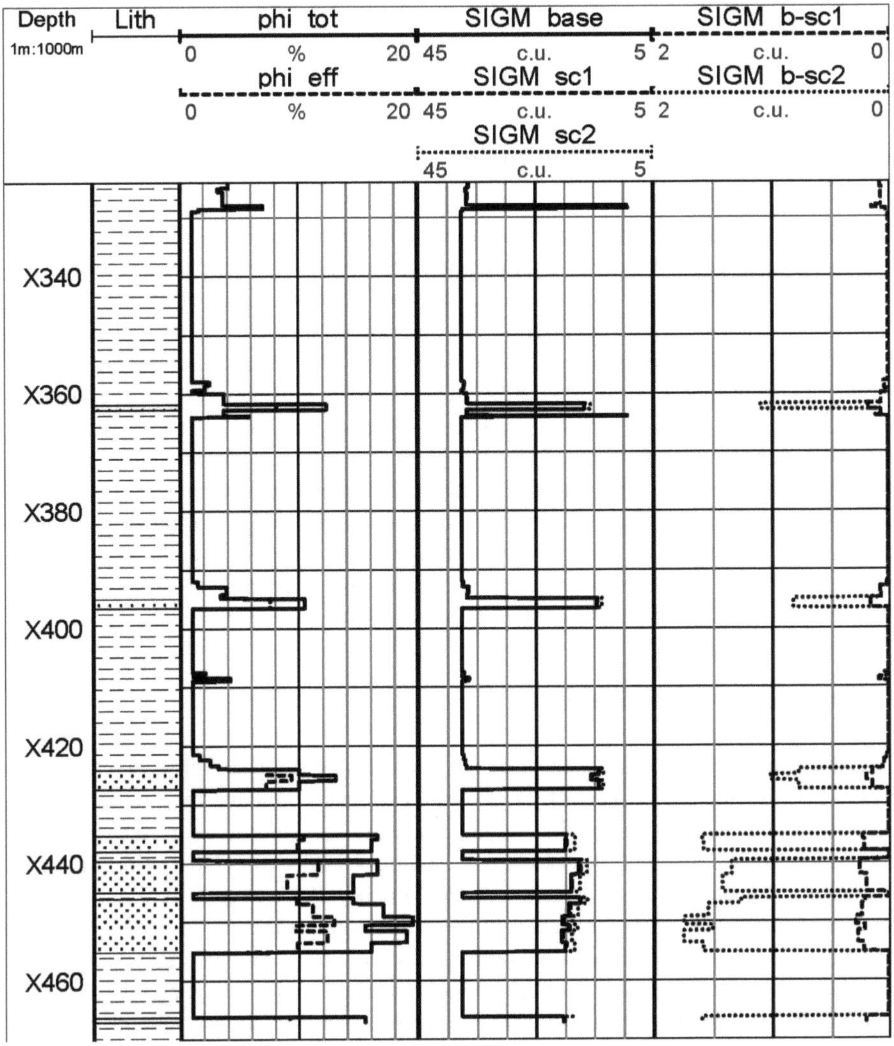

Fig. 5.3 Σ modelling results for the three different scenarios for the S13 injection well at the Altmark site. Injection intervals with high permeabilities are indicated by dotted signature. *Left panel*: total porosity (phi_tot)* and effective porosity (phi_eff)*, *middle panel*: baseline (SIGM_b) and repeat Σ (SIGM_sc1, SIGM_sc2), *right panel*: difference between baseline and repeats (SIGM_b-sc1, SIGM_b-sc2) (Data according to GDF SUEZ 2009)

5.2 Active and Passive Seismic Monitoring

Within the project CLEAN a set of active and passive seismic surveys were designed to monitor the spatial and temporal evolution of the CO_2 front within the reservoir as well as the related processes. These experiments comprised a time-lapse MSP/VSP survey as well as the installation of a borehole seismometer network for monitoring and analysis of potential injection-induced seismicity.

5.2.1 Active Seismic Monitoring

To monitor the spatial and temporal evolution of the injected CO_2 in the subsurface, a time-lapse MSP/VSP survey was planned with well S1 as the observation well, which has an offset of approximately 600 m to the main injector S13. This comprised a baseline survey before and a repeat measurement approximately 1 year after start of injection. In order to design the specific acquisition parameters, i.e. the position of the source locations at the surface and the receiver locations in the well, as well as to estimate the expected variations due to CO_2 injection, the expected changes of rock-physical parameters were estimated and the corresponding seismic wave field simulated. The amplitude changes in the wave fields were estimated by comparing the amplitudes of the seismograms while time shifts were calculated with coda-wave interferometry (Snieder et al. 2002). These calculations are mandatory boundary conditions for a reliable assessment of the detectability and repeatability of the planned time-lapse experiments.

5.2.1.1 Rock Physical Estimates

First the expected changes of the physical properties of the reservoir rock due to the replacement of the reservoir gas by CO_2 were estimated. For that reason the acoustic velocity and density of CO_2 for varying pressure and temperature and the main components of the reservoir gas at reservoir conditions were calculated using equations of state for the particular gas (Setzmann and Wagner 1991; Span and Wagner 1996; Span et al. 2000). From the properties of the single components the total density of the reservoir gas was obtained by calculating the molar average of the single components and the total velocity by determining the molar average of the inverse bulk modulus.

The properties of the gas saturated reservoir rock are then obtained by using these results for the two pore fluids. The density is again calculated by taking the molar average of the rock and the pore fluid into account, whereas the velocity is determined by:

$$v_{p,sat} = \sqrt{\frac{K_{sat} + \frac{4}{3}\mu_{sat}}{\rho_{sat}}} \tag{5.1}$$

with the saturated moduli K_{sat} and μ_{sat} obtained from the Gassmann equations:

$$K_{sat} = \frac{K_0\left(\frac{K_{dry}}{K_0 - K_{dry}} + \frac{K_{fl}}{\phi\left(K_0 - K_{fl}\right)}\right)}{1 + \frac{K_{dry}}{K_0 - K_{dry}} + \frac{K_{fl}}{\phi(K_0 - K_{fl})}}$$

$$\mu_{sat} = \mu_{dry}. \tag{5.2}$$

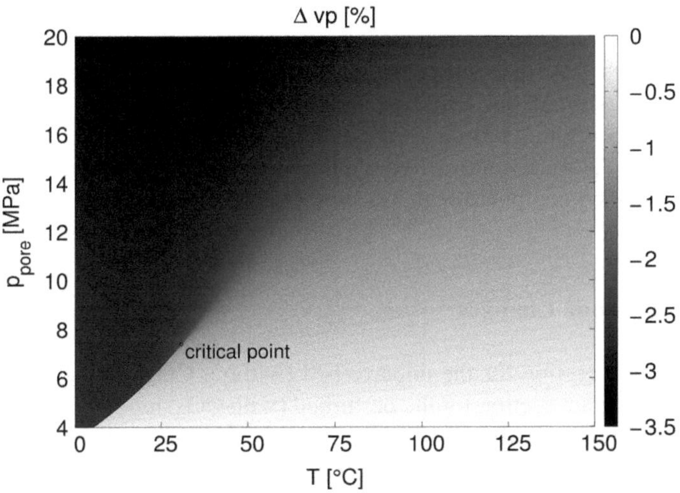

Fig. 5.4 Change in P-wave velocity of the reservoir rock saturated with CO_2 for different pore pressures and temperatures

Here K_{dry} is the dry bulk modulus and μ_{dry} the dry shear modulus, which describe the particular modulus of the dry rock. K_0 is the grain bulk modulus, the bulk modulus of the mineral material making up the rock. K_{fl} is the bulk modulus of the pore fluid, which is the initial gas mixture prior to injection and pure CO_2 after injection. Changes of the rock matrix due to changes in the differential pressure (and consequently changes in pore pressure) were implemented by use of the Eberhart-Phillips-Equation (Zimmermann et al. 1986). The change in P-wave velocity of the saturated rock for different pore pressures and temperatures after injection are shown in Fig. 5.4. These changes are normalized to the particular value of the pre-injection reservoir conditions (saturated with reservoir gas, P = 4.5 MPa, T = 125 °C).

As expected the changes are very small and only for large temperature and pressure changes they become significant. The S-wave velocity changes only due to density and as no S-wave data of the subsurface exists, they were approximated with $v_{s,sat} = v_{p,sat} \cdot 0.58$, which stood the test in many experiments.

5.2.1.2 Seismic Wave Field Simulation

With the information from the rock physical estimates 2D elastic finite-difference modelling was performed (Bohlen 2002) of the corresponding seismic wave field. In a first step a velocity and density model of the subsurface was built based on existing logging information from the monitoring well. In the second step the density and velocity in the reservoir layer was changed (a thickness of 24 m was assumed) within a predefined region of 500 m radius around the injection point.

According to the estimates above a velocity change of −2.14 % and associated density change of 5 % in the reservoir were calculated. These changes reflect an upper boundary of the expected changes and the observed changes during CO_2 injection in the Altmark gas field will most likely be smaller.

For both models the seismic wave field for a certain source location at the surface (offset 5,850 m) and a hypothetical 390-level 3C-receiver array in the borehole were then computed with a distance of 10 m between the receivers.

5.2.1.3 Amplitude Changes

The seismogram section for the unperturbed (without CO_2) model was subtracted from the seismogram section for the perturbed (with CO_2) model (normalized to the maximum amplitude of the dominant period of the source wavelet). Figure 5.5 shows these difference sections for source offsets of 2,900 m (a and b) and 5,850 m (c and d).

For an offset of 2,900 m a pronounced change in the converted S-wave in the horizontal component is observed (Fig. 5.5a) and additionally to S-wave effects a smaller change in the directly reflected P-wave in the vertical component (Fig. 5.5b). The maximum values of amplitude changes are on the order of 5%–10 % and the depth of the corresponding receivers can be identified. For a larger offset of 5,850 m the amplitude changes of the directly reflected P-wave become more significant, while the effects of the converted S-wave diminish. Here, the corresponding values are up to 15 % (Fig. 5.5c) and can be observed in a much smaller depth range (X900 – X400m) than for the smaller offset.

5.2.1.4 Coda-Wave Interferometry

Beside amplitude changes particularly time shifts resulting from velocity changes alone can be expected. They were estimated with coda-wave interferometry, which was proposed by Snieder et al. (2002). Here the changes in the wave fields were quantified by calculating the time shifted cross-correlation of the unperturbed ($u_u(t)$) and perturbed ($u_p(t)$) wave field over a time window $2t_w$ at the centre time t

$$R(t_s) = \frac{\int_{t-t_w}^{t+t_w} u_u(t')u_p(t'+t_s)dt'}{\sqrt{\int_{t-t_w}^{t+t_w} u_u^2(t')dt' \int_{t-t_w}^{t+t_w} u_p^2(t')dt'}} \tag{5.3}$$

where t_s denotes the time shift between the two wave fields. The correlation coefficient $R(t_s)$ attains its maximum for the time shift where perturbed and unperturbed waveforms are most identical. E.g. if the perturbed waveform is a time shifted version of the unperturbed waveform ($u_p(t) = u_u(t-t_p)$), then $R(t_s)$ attains its maximum $R_{max}(t_{max})$, if $t_{max} = t_s = t_p$.

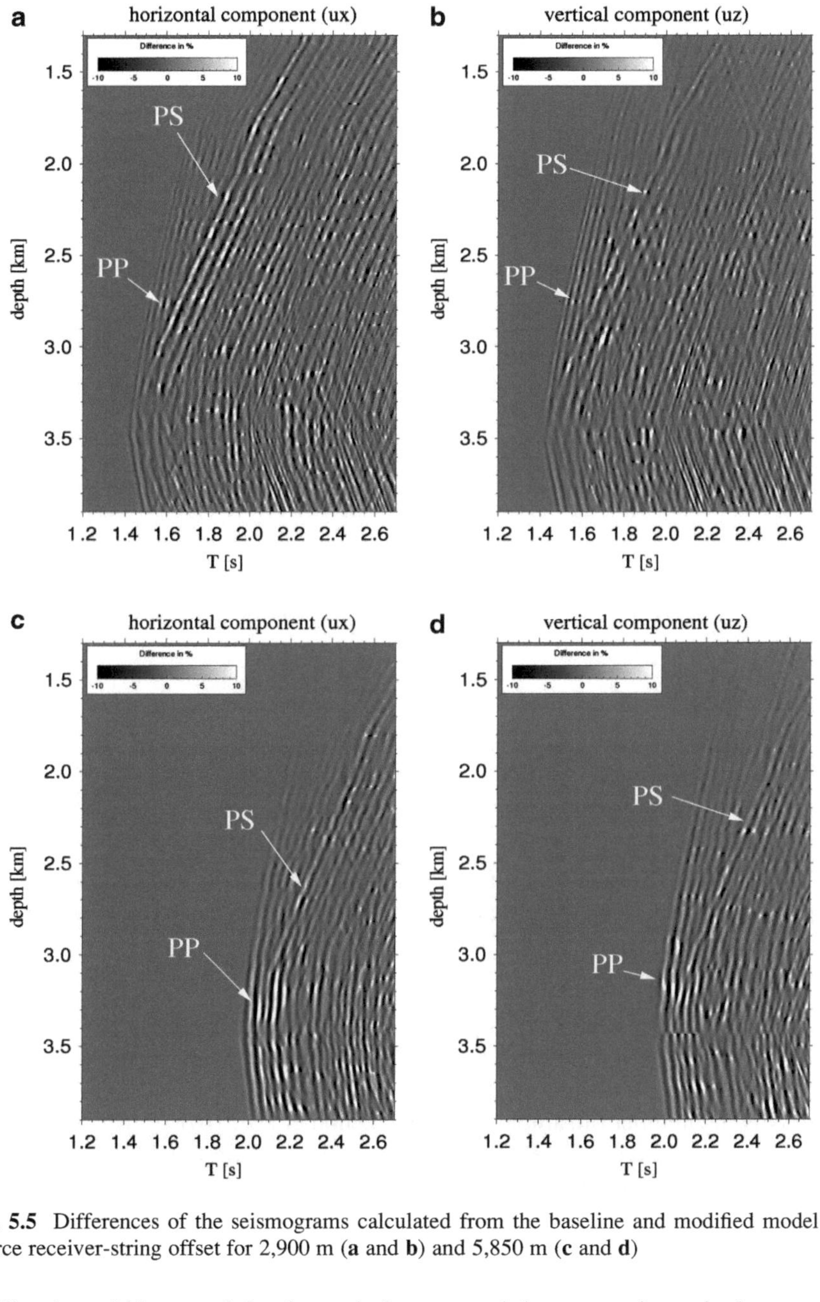

Fig. 5.5 Differences of the seismograms calculated from the baseline and modified model for source receiver-string offset for 2,900 m (**a** and **b**) and 5,850 m (**c** and **d**)

The time shift t_{max} of the time window around the centre time t is the average perturbation of the travel time of all wavelets travelling along all possible trajectories and arriving in the employed time window. Consequently the time shift is correlated with the relative velocity change by

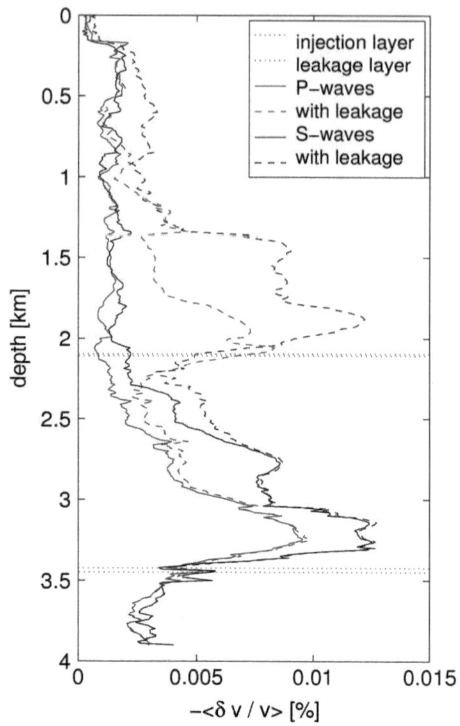

Fig. 5.6 Calculated mean velocity changes for P- (*grey*) and S-waves (*black*) for models with velocity changes in the reservoir only (*solid*) and in an additional leakage layer (*dashed*)

$$\frac{\delta v}{v} = -\frac{t_{max}}{t} \tag{5.4}$$

This was done for each centre time in the seismic sections and finally the mean velocity change over one trace (and due to the VSP arrangement over depth) was calculated with

$$\left\langle \frac{\delta v}{v} \right\rangle = \frac{1}{N} \sum_{i=1}^{N} -\frac{t_{max,i}}{t_i} \tag{5.5}$$

where N is the number of samples per trace (Zhou et al. 2010).
Figure 5.6 shows the mean velocity changes in the P- (grey) and S-waves (black) calculated for the model described above. Largest mean velocity changes are obtained above the reservoir with a maximum of about 0.01 % for P-wave, which is assumed to be the threshold. The maximum for S-waves is even higher. An additional leakage layer of 10 m thickness was implemented into the model and a velocity decrease of −4 % received, which is a lower boundary for CO_2 replacing water (Xue and Ohsumi 2004). The results (dashed curves) show effects of about 0.01 %.

5.2.2 *Passive Seismic Monitoring*

The location of potentially induced seismic events during CO_2 injection is important for different reasons, e.g. on the one hand for the identification of pre-existent faults and to detect potential leakages (Rudledge 2009) and on the other hand to differentiate between induced and natural seismicity. Locating such events manually represents a great challenge, due to their potentially large number and low magnitude. In the frame of the project CLEAN a semi-automatic location procedure based on the diffraction-summation method was developed to enable the location of a large number of low-magnitude seismic events almost automatically and in real-time without picking. Similar successful approaches based on reverse time migration have been already published by Gajewski and Tessmer (2005).

5.2.2.1 Method

The used locating approach is based on the application of the diffraction-summation method, which is routinely used for migration of active-source seismic data sets. For this method the subsurface is covered with a Cartesian grid. Each of these grid points is a potential hypocentre. For a given velocity model, the theoretical travel times between the receivers and each subsurface point are calculated. Then, the seismogram amplitudes are summed up along the corresponding diffraction surfaces and the sum is assigned to the respective subsurface grid point. This procedure is repeated for every potential origin time and a 3D image volume is obtained for each of these origin times (Fig. 5.7). The absolute spatial and temporal maximum of all imaged volumes provides the hypocentre and the true origin time.

5.2.2.2 Application to Synthetic Data

First, the location method was tested on synthetic datasets consisting of homogeneous acoustic velocity models and explosion point sources. In addition complex elastic velocity models based on the geology of the Altmark natural gas field as well as a P- and S-wave emitting single force were implemented. Furthermore a double couple source shall be implemented. The network geometry was adapted to the planned receiver geometry in the Altmark area (Fig. 5.8). There, the network is supposed to consist of three shallow boreholes with depths of 200 m each and equipped with five three component geophones at depths between 80 m and 200 m with an interval of 30 m. Additionally, a deep observation well equipped with five geophones at depths between 3,100 m and 3,140 m with distances of 10 m between each geophone is part of the seismological network. The synthetic seismograms were calculated with Finite Differences (Bohlen 2002) for a dominant event frequency of 50 Hz. Before the location of the seismic events, processing of the seismic traces was necessary. This comprised the muting of all onsets after the P-wave to eliminate e.g. the S-wave, which has generally higher amplitudes and which would dominate the result of the

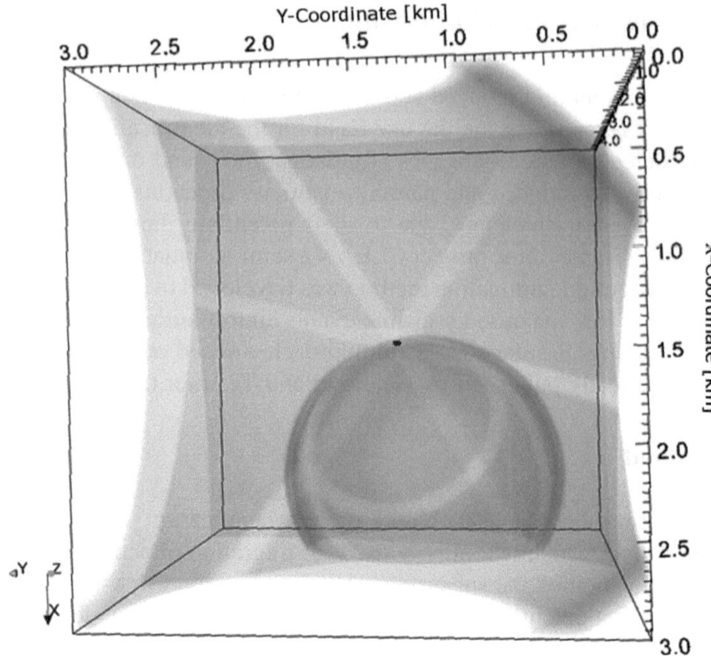

Fig. 5.7 Iso-amplitude representation of an exemplary image volume for the true origin time. The maximum in the centre of the image volume denotes the determined location of the hypocentre

location. Furthermore every single seismic trace was normalized to the maximum of the absolute value and its envelope was calculated. Figure 5.9 shows a representative successful location result; the black point marks the hypocentre determined with the diffraction-summation method and the white point denotes the true hypocentre.

These location tests using synthetic data show that the method in its current implementation is well suited for the efficient and reliable location of seismic events using sparse mixed surface and borehole receiver networks.

5.2.2.3 Network Sensitivity

Whether a failure in the cap rock can be seismologically detected, depends on the sensitivity of the seismological network. With known network sensitivity, it is conversely possible to decide, whether a rupture process with certain strength did occur or not.

For the seismological network in the Altmark the calculations of the detectability show, as expected, that the highest sensitivity is achieved in the centre of the network. It is there possible to record seismic events in a depth of 3,000 m with moment magnitudes Mw $= -0.5$ on all 20 receivers (Fig. 5.10, top). The receivers in the deep observation well can detect seismic events with moment magnitudes Mw $= -2$ in the vicinity of the injection well (Fig. 5.10, bottom).

Fig. 5.8 Map of the seismological network in the Altmark area

5.3 Stable Isotope and Geochemical Monitoring

Gas and fluid monitoring based on stable carbon ($\delta^{13}C$) and oxygen ($\delta^{18}O$) isotopes are established methods in aquatic and atmospheric geosciences. Adaption to Enhanced Gas Recovery (EGR), Enhanced Oil Recovery (EOR) and Carbon Capture and Storage (CCS) require careful consideration of the influences of high p/T conditions and the increased salinity of deep geochemical fluid systems on the carbonate equilibrium as well as on the isotope fractionation of the carbon species.

The delay of CO_2 injection during the CLEAN project opened the opportunity of focus fieldwork to baseline monitoring of DIC (dissolved inorganic carbon) concentrations, CO_2 pressures (pCO_2), $\delta^{13}C$ and $\delta^{18}O$. These parameters were interpreted based on a theoretical study of the influences of the ambient conditions of deep formations. Furthermore, they were compared to an extensive set of laboratory experiments on stable isotope fractionation, partitioning and related geochemical processes under these extreme conditions. Additionally, fields of future research were identified to further extend the knowledge gained in this project to other types of carbon storage reservoirs at different temperature and pressure regimes as well as geochemical conditions.

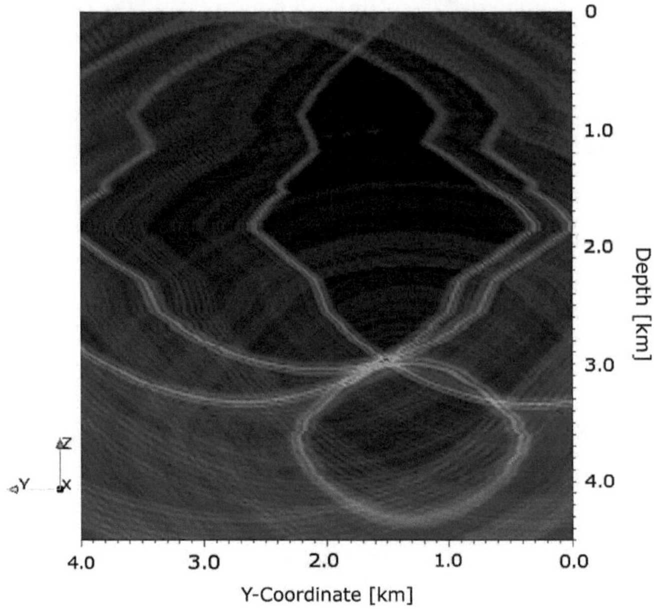

Fig. 5.9 Vertical depth slice through the image volume for the complex Altmark model. The *black point* denotes the hypocentre determined using our diffraction-summation approach, the white point denotes the true hypocentre

5.3.1 Baseline Monitoring

For monitoring reservoir baseline conditions with respect to DIC, $\delta^{13}C$ and $\delta^{18}O$, seven sampling campaigns were carried out in the course of the CLEAN project. During these campaigns, fluids from eight boreholes were sampled and deep samples acquired from three different boreholes as listed in Table 5.2.

Reservoir fluids were sampled using either an open flow-through sampler ('Doppelkugelbüchse', DKB), or a pressure-sealed Permanent Displacement Sampler (PDS). Influences of these different sampling techniques on DIC concentration and isotope ratios are discussed in Myrttinen et al. (2010a). In general, differences depend on the pressure and pCO_2 gradient between surface and sampling depth. Thus, for baseline monitoring with its low and rather constant DIC concentrations from 1 to 3 mmol C L^{-1}, both samplers may be applied. However, after injection and increase by orders of magnitude of pCO_2, the PDS should be preferred to obtain accurate samples.

The mean $\delta^{13}C_{DIC}$ and $\delta^{18}O_{H_2O}$ values as well as their standard deviations from the total of 29 depth samples collected between 3,000 and 3,515 m below ground are given in Table 5.2. Most of the measured values are in good agreement to the

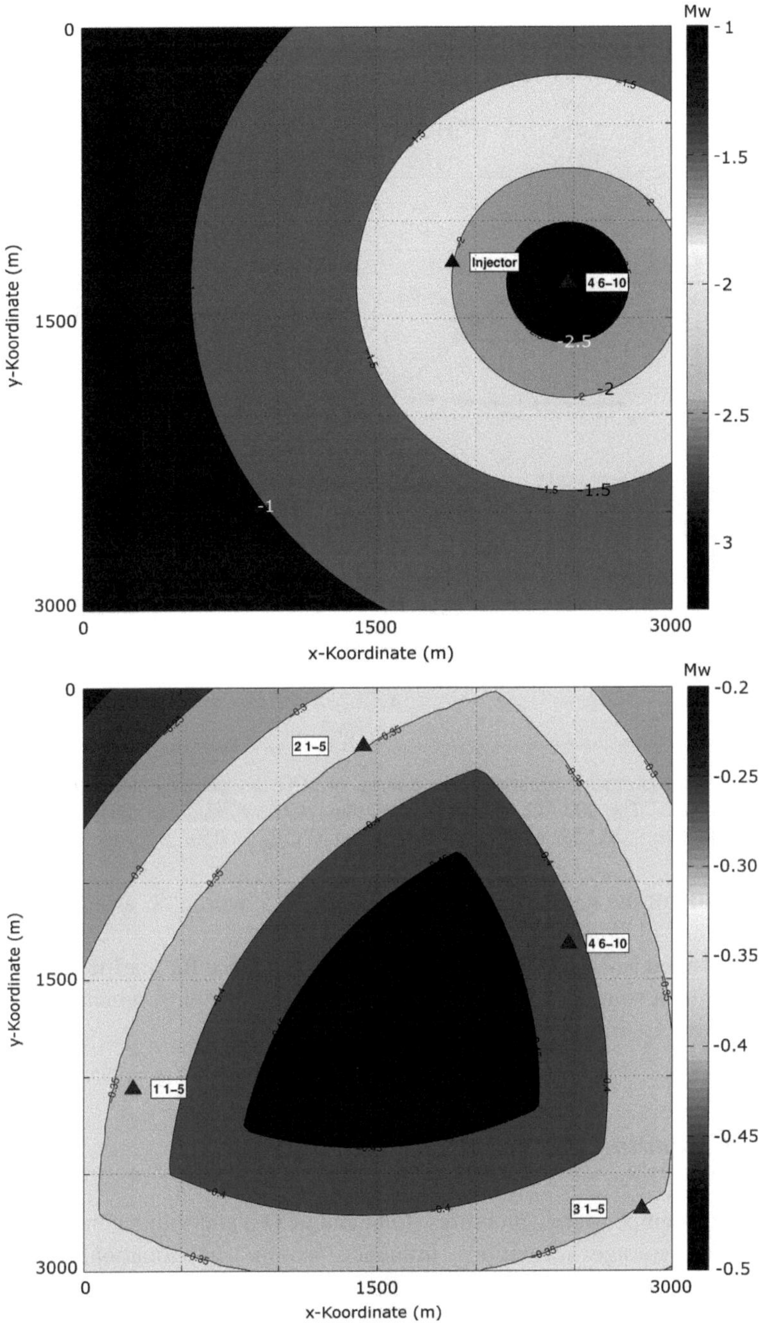

Fig. 5.10 Network sensitivity of all receivers (*top*) and deep receivers (*bottom*)

Table 5.2 Baseline monitoring data sampled during the project

Medium	Borehole	Depth (m)	Date	$\delta^{13}C$ ‰	$SD_{\delta 13C}$	$\delta^{18}O$ ‰	$SD_{\delta 18O}$
Rock	S75	3,366–3,410	–	−3,1	±0,57	−8,7	±0,83
Fluid	S10	3,336	11.2010	−4,1	±0,32	−3,3	±0,02
	S13	3,420	03.2009	−11,5	–	−6,3	–
	S17	3,436	04.09–10.10	−2,0	±1,47	6,3	±2,41

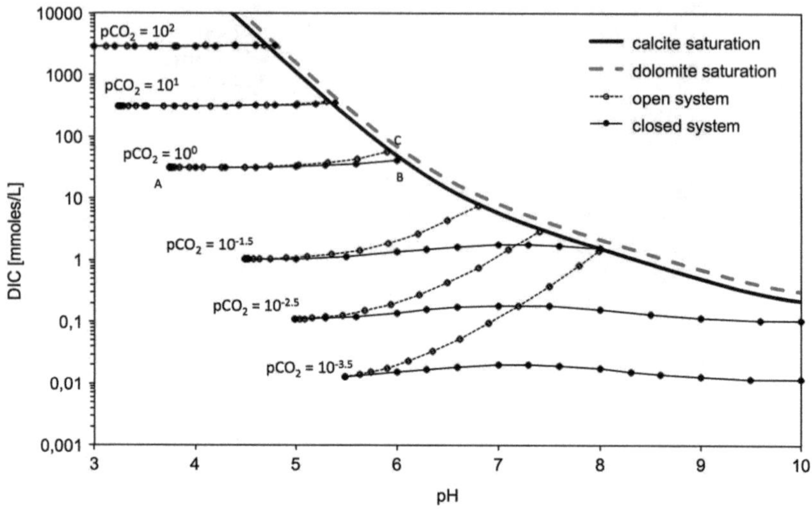

Fig. 5.11 DIC-concentration against pH for a range of pCO_2 in open and closed systems. TDS = 6 mol kg^{-1} NaCl, T = 200 °C, P = 20.0 MPa plus pCO_2. $\delta^{13}C_{DIC}$ for point A = −28.6 ‰; B = −19.4 ‰; C = −30.1 ‰. $\delta^{13}C_{CO2}$ = −28 ‰ and $\delta^{13}C_{Calc}$ = 0 ‰

isotope values of the representative rock sample S75, with $\delta^{13}C$ around −3 ‰ and $\delta^{18}O$ between −3 ‰ and −7 ‰.

The measured isotope ratios are interpreted as evidence for marine carbonates as the predominant source of the DIC within the reservoir fluids, which is consistent with data from Lüders et al. (2010).

5.3.2 Extending the Theory

Because high temperatures, pressures (total and pCO_2) and solute concentrations of deep carbon storage formations influence isotope fractionation, equilibrium constants and kinetics of the carbonate equilibrium, an extended approach for CO_2 isotope investigations is required. Known dependencies of the governing equations of CO_2 dissolution, DIC speciation and $\delta^{13}C$ fractionation were combined to enable the transfer of well-tested isotope methods from near-surface areas to deep formations (Becker et al. 2011).

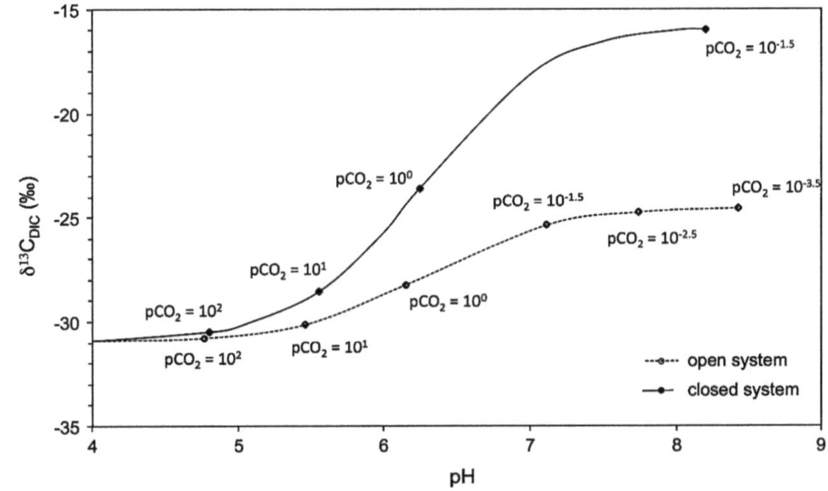

Fig. 5.12 $\delta^{13}C_{DIC}$ progression in open and closed systems at 50 °C at various pCO_2. The points indicate values at calcite saturation for each pCO_2. End-member $\delta^{13}C$ is assumed to be −30 ‰ for injected CO_2 and of 0 ‰ for soil carbonates

For convenient evaluation of the ambient conditions of water samples gained from injection sites, a DIC/$\delta^{13}C_{DIC}$ scheme for typical conditions of CO_2 storage sites were developed for the expected ambient conditions (Fig. 5.11). Reservoir conditions considered for calculating this scheme reach reservoir pressures of up to 20 MPa, pH down to 3 temperatures of up to 200 °C and pCO_2 of up to 10 MPa.

For the calculation of the DIC speciation, the constants K_1* and K_2* of Millero et al. (2007), including activity corrections for ionic strength and temperature, were applied. A temperature-dependent K_{CO2} (Drever 1997) was modified with fugacity coefficients from Duan et al. (1992) to account for non-ideal behaviour of the CO_2 depending on pressure and temperature and gas phase composition. The scheme differentiates between open or closed conditions with respect to CO_2 supply as these may lead to different equilibrium conditions and may change repeatedly during and after CO_2 injection (Fig. 5.12). In case of open system dissolution, $\delta^{13}C$ was calculated as

$$\delta^{13}C_{DIC} = \frac{\left(\begin{array}{c}(H_2CO_3)\cdot\left(\delta^{13}C_{CO_2}+10^3\ln\alpha_{CO_2-H_2CO_3}\right)+\left(HCO_3^-\right)\\ \cdot\left(\delta^{13}C_{CO_2}+10^3\ln\alpha_{CO_2-HCO_3^-}\right)\\ +\left(CO_3^{2-}\right)\cdot\left(\delta^{13}C_{CO_2}+10^3\ln\alpha_{CO_2-CO_3^{2-}}\right)\end{array}\right)}{(H_2CO_3)+\left(HCO_3^-\right)+\left(CO_3^{2-}\right)} \tag{5.6}$$

and for closed systems as

$$\delta^{13}C_{DIC} = \frac{\begin{pmatrix} (H_2CO_3)_{ini} \cdot \left(\delta^{13}C_{CO_2} + 10^3 \ln \alpha_{CO_2-H_2CO_3}\right) + \left(HCO_3^-\right) \\ \cdot \left(\delta^{13}C_{CO_2} + 10^3 \ln \alpha_{CO_2-HCO_3^-}\right) \\ + \left(CO_3^{2-}\right) \cdot \left(\delta^{13}C_{CO_2} + 10^3 \ln \alpha_{CO_2-CO_3^{2-}}\right) + (CaCO_3) \cdot \delta^{13}C_{calcite} \end{pmatrix}}{(H_2CO_3) + \left(HCO_3^-\right) + \left(CO_3^{2-}\right) + (CaCO_3)_{diss}} \tag{5.7}$$

with 10^3 lnα being the isotope fractionation factors between species given in the respective indices. Equations 5.6 and 5.7 are extended versions of schemes developed by Clark and Fritz (1997). The lowest pH values for each pCO_2 then result from equilibrium DIC speciation without dissolution of $CaCO_3$. Higher pH values are obtained by increasing dissolution of $CaCO_3$. Background concentrations are assumed to be pure NaCl to reduce complexity, without introducing significant errors (Kervévan et al. 2005).

The $\delta^{13}C_{CO_2}$ was assumed to be −28 ‰ and the $\delta^{13}C_{CaCO_3}$ as 0 ‰. These values can easily be adapted to real values measured from injected CO_2 and usually $CaCO_3$ dominated baseline values. The large amount of dissolved CO_2 at $pCO_2 > 1$ MPa leads to a diminishing influence of the baseline $\delta^{13}C_{DIC}$ on the resulting isotope ratio and also lowers the pH and thus renders DIC to consist mostly of CO_2(aq). Hence, the limiting value of the $\delta^{13}C_{DIC}$ towards low pH values is given by the $\delta^{13}C_{co_2}$ modified by the isotope fractionation between CO_2(g) or supercritical CO_2 (i.e. CO_2(sc)) and CO_2(aq). This constant $\delta^{13}C_{DIC}$ values can serve as solid end-members for further mixing calculations to monitor the migration of the trapped CO_2.

Isotope balances of $\delta^{13}C_{DIC}$ alone will therefore be less precise than in near-surface surroundings, due to the reduced sensitivity of these solid end-members to mixing with carbon from minor sources. Therefore, laboratory experiments have been designed and conducted to extend knowledge about temperature and pressure dependence of ^{13}C fractionation for most precise determination of fluid mixing. Furthermore, $\delta^{18}O$ as an indicator with improved sensitivity to fluctuations at high pCO_2 have been established. Experimental setup for both isotope methods is explained in the following section.

5.3.3 Laboratory Experiments

Laboratory experiments using typical baseline and CO_2 values were conducted to investigate isotope fractionation and shifts of $\delta^{13}C_{DIC}$ and $\delta^{18}O_{H2O}$ under various pressures, temperatures and baseline DIC concentrations.

5.3.3.1 $\delta^{13}C_{DIC}$

The experimental setup was based on a temperature-pressure device consisting of a 9 mL high-pressure autoclave, connected to a CO_2 tank (Fig. 5.13, left).

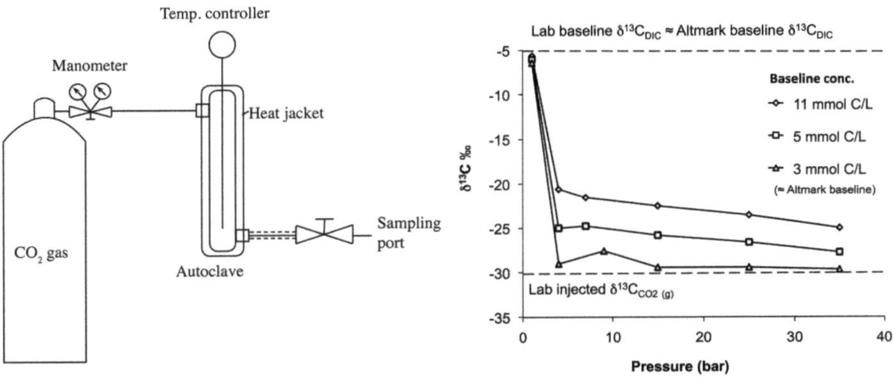

Fig. 5.13 Schematic of the temperature–pressure device (*left*). $\delta^{13}C_{DIC}$ shifts with varying pressures (*right*)

The autoclave including the attached tubing was wrapped with a digitally controlled heater. Pressure was controlled with a manometer. The following results (Fig. 5.13, right) apply to experiments (expanded after Myrttinen et al. 2010a) conducted with baseline DIC concentrations of 3, 5 and 11 mmol L^{-1}, respectively, at 60 °C and pCO_2 ranging between 0.4 MPa and 4.5 MPa. An increase in pCO_2 causes a negative $\delta^{13}C_{DIC}$ isotope shift due to the addition of the more negative CO_2(g). A higher DIC baseline value has a buffering effect on the negative shift. The DIC concentration increases with pCO_2 due to the forced dissolution of additional CO_2.

5.3.3.2 $\delta^{18}O_{H_2O}$

On the one hand $\delta^{13}C$ usually responds rapidly to very small traces of CO_2 at the tip of CO_2 plumes (Myrttinen et al. 2010b) and therefore is an ideal tracer for leakage monitoring or early detection of a spreading CO_2 plume. On-going studies on $\delta^{18}O_{H_2O}$, on the other hand, show that a noticeable isotope shift can only be expected if contact exists to very large amounts of CO_2, as they occur during CO_2 injections (Johnson and Mayer 2011). The high amount of oxygen in water consequently equilibrates with the oxygen in CO_2 and causes a sort of memory effect, allowing the detection of exchange between both phases even after the contact to the CO_2 has ended. Therefore, $\delta^{18}O$ of both CO_2 and water can serve as a tracer for residence times and propagation velocities of highly pressured CO_2 plumes as well as mass ratios of CO_2(g) and water. Only preliminary results have been published so far (Johnson and Mayer 2011), requiring in depth investigations on kinetics and sensitivity of $\delta^{18}O_{H_2O}$ to DIC changes.

Within the CLEAN project, kinetics of $^{18}O_{H_2O}$ equilibration with high volumes of CO_2 were investigated for different salinity conditions. First results showed that the molal ratio of CO_2 and water determines the $\delta^{18}O$ shift. Therefore, experiments could be conducted at ambient pressures at high CO_2/H_2O ratios. For this, CO_2 gas

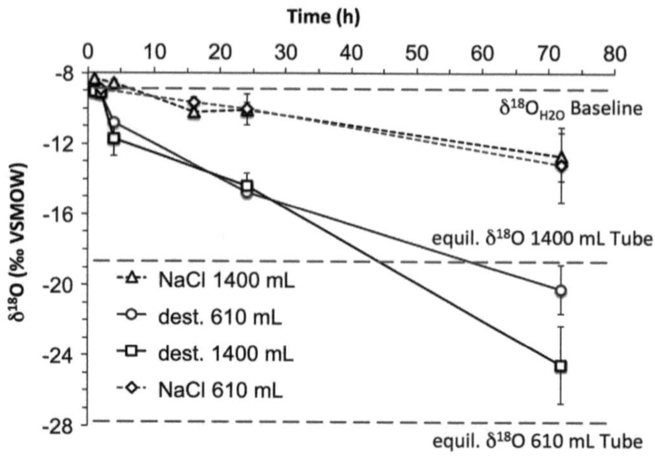

Fig. 5.14 Equilibration kinetics of $\delta^{18}O$ for different gas/fluid ratios with fresh and NaCl-saturated water

was brought into contact to water in gas sampling tubes of two different sizes (610 mL and 1,400 mL) for various time intervals. Oxygen ratios of CO_2 and water were 1.25:1 and 2.88:1 respectively. Determined equilibrium values of $\delta^{18}O_{H_2O}$ are shown in Fig. 5.14. Two series were conducted with fresh water and then repeated with highly saline, NaCl saturated water (~6 molal). Samples were shaken constantly to increase the CO_2 exchange via the gas/water interface to mimic turbulent mixing between water and CO_2.

The effect of salinity on equilibration kinetics is shown in Fig. 5.14. They show that higher salinity hamper the isotope exchange between CO_2 and water. While $\delta^{18}O$ shifts in both phases are highly promising, processes need to be studied in further detail to increase the investigated intervals of ambient conditions. This will help to formulate general advection-dependent relations for equilibration kinetics and isotope fractionation. Nonetheless, a good base of knowledge about $\delta^{18}O$ methods in carbon dioxide storage was established here.

5.4 Biogeochemical and Microbial Characterization of Reservoir Fluids

Potential geological reservoirs for CO_2 (e.g. saline aquifers, hydrocarbon reservoirs) contain complex biogeochemical systems, including a broad diversity of minerals, brines and largely uncharacterized microbial ecosystems. Thorough studies of samples from deep boreholes, using a variety of molecular techniques, showed an active biosphere composed of diverse groups of microorganisms which play an important role in the transformation of material within the earth crust (Parkes et al. 2000; D'Hondt et al. 2004; Sahl et al. 2008). In these extreme environments well-adapted and highly active microorganisms exist, and play an important role in reservoir biogeochemical cycling (Aloisi et al. 2002).

Many important processes in geological systems are partly or even completely catalysed by microorganisms.

The injection of large volumes of CO_2 influences the environmental conditions especially pH, temperature and pressure and may induce re-arrangement und relocation of minerals and organic substances. These changes in reservoir conditions can stimulate microbial activity, which might have an impact on the subsurface rock material. For example, microbial activity due to CO_2 injection might enhance the dissolution of certain minerals, thus increasing the permeability and affecting the storage capacity. Conversely, the precipitation of mineral or biological compounds may lead to a loss of injectivity in the well near area (Morozova et al. 2010). Microbial metabolic activity influences the global material cycle and has led to significant mineral deposits in the geological time frame (Welch and McPhail 2003). The precipitation of iron sulphide, silicate, calcite and dolomite minerals or the dissolution of feldspar and Fe (III) minerals are examples on how bacteria can influence mineralogical characteristics (Douglas and Beveridge 1998; Hutchens et al. 2003; Sanchez-Roman et al. 2007; Vorhies and Gaines 2009). Furthermore, reactions between supercritical CO_2 and rock materials might produce fluids with increased concentrations of dissolved organic compounds, which can supply the autochthonous microorganisms with energy.

In addition, the growth of microorganisms on the metal surface (biofilms) can have a profound effect on metal deterioration, known as microbially-influenced corrosion (MIC, Beech and Sunner (2004)). Microorganisms are known to induce and/or speed up corrosion processes by producing organic and inorganic acids and ammonia, which lead to metal surface ionization (Videla and Herrera 2005). Given the complexity of the MIC mechanisms, the processes of the biocorrosion are still poorly understood (Beech and Sunner 2004; Little and Lee 2007; Javaherdashti 2008).

Within the CLEAN project, the microbial community of the natural gas reservoir was investigated in order to assess their interaction with reservoir components and CO_2 storage. Since microorganisms represent very effective geochemical catalysts, the thorough investigation of their distribution and physiology could be important for the process of CO_2 storage. This study characterized the microbial community of the deep biosphere of the Altmark reservoir by 16S rRNA gene sequence analysis and cell quantification. Fluid samples from three wells, S10, S13 and S17differing in their temperature profiles and operative history were investigated.

5.4.1 Sampling

Fluid samples were collected from the reservoir using downhole sampling (Erdöl-Erdgas Workover GmbH, Fig. 5.15a) at perforation depths in S17 and S13 using double ball-lining (in German “Doppelkugelbüchse”, DKB) and bottom hole positive displacement samplers (PDS). The DKB samplers were inserted open into the well, and lowered until the required sampling depth. The DKB was then flushed by well fluids, closed and raised to the surface. The PDS sampler is inserted closed into

Fig. 5.15 Sampling of reservoir fluids using depth sampling technologies (**a**). Heating jacket with PDS sampler inside (**b**). Sampling of reservoir fluids from DKB (**c**) and PDS (**d**) samplers

the well until defined depth. It then opens to take a sample, closes again and is raised to the surface. The PDS sampler was sterilized by heating at 180 °C for 2 h inside the heating jacket (LeutertTM), specially developed for preheating PDS sampler (Fig. 5.15b).

The DKB sampler was cleaned in the laboratory and washed with sterilised deionised water and ethanol right before sampling. The pH, conductivity, temperature and other parameters were measured directly after the sampling process. The fluids were transferred aseptically into sterilised 100–1,000 mL glass vials, cooled to 4 °C and immediately transferred to the laboratory for microbiological analyses. For the geochemical analyses, fluids were collected into sterilised and anaerobically prepared 100–1,000 mL glass vials, which were flushed with N_2 in order to provide anaerobic conditions and kept at room temperature until analysed. Importantly, a nitric acid treatment and water flush was performed prior to sampling in well S13 due to build-up of sand deposits resulting in well clogging. About 4 L nitric acid and 4 m^3 water were added, and 0.4 m^3 fluids were back-produced prior to sampling in order to remove salts and rust from the well. Well S10 was treated with hydrochloric acid (during 3 months, about 2–8 m^3 per month) and tenside (surface-active agent) about 10 years prior to sampling but was not operated since then. Well S17 was treated with hydrochloric acid (1 m^3, about three times during 1 year) 20 years prior to sampling. No tenside treatment was performed. This well was in operation until about 1 year prior to sampling.

5.4.2 Geochemical Analyses

Electrical conductivity, pH and fluid temperature were measured at the surface using a portable pH/mV/temperature meter (WTW). The total organic carbon content was determined using a TOC-analyser (Total Organic Carbon, Dimatec

GmbH) according to DIN EN 1484-H3. Main and trace elements in fluids were measured optically with inductively coupled plasma atomic emission spectroscopy (ICP-AES). The isotopic ratio of gases (CO_2, CH_4) was determined with gas chromatography isotope ratio mass spectrometry (GC-IRMS).

5.4.3 Molecular Approaches Applied to the Reservoir Fluids

Molecular fingerprinting is a tool for determination of microbial diversity in environmental samples without requiring cultivation of the detected species (Schwieger and Tebbe 1998). The approach bases on the amplification of 16S rRNA gene fragments from DNA directly extracted from the environment and the subsequent electrophoretical separation of the PCR products on a polyacrylamide gel (SSCP, single-strand conformation polymorphism) or on a gradient gel (DGGE, denaturating gradient gel electrophoresis) resulting in a genetic fingerprint of the dominant community structure. For DNA extraction, microbial cells were concentrated by filtration of the reservoir fluids on 0.2 μm filter units (Millipore). Nucleic acids were extracted from preserved filters with Ultra Clean DNA Isolation Kit (Mo Bio Laboratories) according to manufacturer's suggested protocols with few modifications. 16S rRNA subunits were amplified by polymerase chain reaction (PCR) and nested PCR using different bacteria and archaea-specific primers. PCR products were analysed by electrophoresis via SSCP (Orita et al. 1989) and DGGE techniques (Muyzer et al. 1993). The obtained 16S rRNA gene sequences were compared with the sequences available in the GenBank database using BLAST (http://blast.ncbi.nlm.nih.gov/Blast.cgi).

For the quantification of the microbial community different techniques were used. The dye SYBR Green binds to all DNA of living and dead cells and was used to determine the total cell number (Weinbauer et al. 1998), modified. Additionally, the method catalysed reporter deposition fluorescence in situ hybridization (CARD-FISH) was used for analyses (Pernthaler et al. 2002). Probes used for CARD-FISH target quickly degradable rRNA in bacteria and archaea and, therefore, detect only living cells. For analysing the reservoir fluids, sample volumes of 1–100 mL were filtered to determine cell numbers with SYBR Green direct counting and CARD-FISH. The detection limit for both microscopic methods was around 1,000 cells per mL. Another method used, the quantitative polymerase chain reaction (qPCR), quantifies amplifiable DNA presumably of living, and to some extent dead cells (Schippers and Neretin 2006). QPCR can be used to quantify dependent on the primers 16S rRNA genes of e.g. bacteria, archaea or geobacteraceae or to quantify the18S rRNA genes of eukarya. Also specific functional genes, like dsrA and mcrA, can be quantified with this method. The DNA copy numbers can be converted to cell numbers with conversion factors from a ribosomal RNA operon copy number database. The detection limits for qPCR analyses were 100–1,000 DNA copies per mL.

5.5 Results and Discussion

The Altmark reservoir represents an extreme environment for microbial life with temperatures between 120 °C and 130 °C (Horikoshi and Grant 1998; Rothschild and Mancinelli 2001; Pikuta et al. 2007). In the reservoir fluids high concentrations of salts up to 400 g per litre and a density of up to 1.23 g per mL were detected, which are typical values for fluids from greater depths (Table 5.3 and Fig. 5.16). Sulphate and TOC (total organic carbon) -concentrations, both relevant for microbial processes, ranged between 400–1,800 mg L^{-1} and 70–428 mg L^{-1}, respectively.

The carbon isotopes in methane outgassing from the reservoir fluids were analysed and showed $\delta^{13}C$-isotopic values between −20 ‰ and −25 ‰ versus PDB typically measured for thermogenic methane of reservoirs.

High temperature is postulated to be one of the most limiting factors controlling microbial growth and survival at great depths (Parkes et al. 2000; Wilhelms et al. 2001; Jorgensen and Boetius 2007). However, results of the Altmark baseline survey indicated the presence of microorganisms in this deep hot subsurface environment. The sequence analyses revealed the presence of several H_2-oxidising bacteria (hydrogenophaga sp., adicdovorax sp., ralstonia sp., pseudomonas sp.) in all fluid samples, thiosulphate-oxidising bacteria (diaphorobacter sp.) in the fluids from the well S17 and biocorrosive thermophilic microorganisms, which have not previously been cultivated, in the fluids from S13 and S17 wells (Fig. 5.17). Furthermore, several sequences of uncultivated microorganisms were found, that were similar to representatives from other saline, hot, anoxic, deep environments (Aragno and Schlegel 1992; Marchesi et al. 2001; Konstantinidis et al. 2003; Duncan et al. 2009). Although well S13 was treated with acid and water prior to sampling, no sequences were identified yet, that clearly indicated a contamination with non-indigenous microorganisms.

For the reservoir fluids from the well S10, comparable fingerprinting analyses of the fluids collected with DKB and sterilized by heating PDS samplers were performed. Our results revealed a similar microbial community composition of the fluids with both sampling methods. This allows the assumption that we were able to characterise the autochthonous microbial community from the deep hot Altmark reservoir and effects of contamination by the applied sampling method were negligible. Similar to the wells S17 and S13, the sequence analyses revealed representatives of H_2-oxidising bacteria (ralstonia sp., pseudomonas sp.); also dissimilatory metal reducers (pantoea sp.), aromatic-degrading deep sediment inhabitants, known to be involved in metal corrosion (sphingomonas sp.) and extremophilic bacteria (bacillus sp.) in reservoir fluids.

Results of the fingerprinting analyses support the assumption that identified microorganisms originate from Altmark deep hot reservoir. Thus, ralstonia, prevotella and pseudomonas species were already found in deep sediments (Marchesi et al. 2001). Together with hydrogenophaga those species are characterized by chemolithotrophic growth (Aragno and Schlegel 1992). Among those bacteria thermophilic species are known (Stohr et al. 2001). Biocorrosive

Table 5.3 Geochemical analysis of the downhole fluid samples for the well S10 and S17 (Geochemical analyses for the fluids from the well S13 are not presented here due to the possible influence of the well treatment prior sampling on geochemical data). Fluid temperature was determined using PT-log prior to sampling

Well	Temperature (°C)	pH	Salinity (g/L)	DOC (mg/L)
S17	120	5.7	395	123.1
S17	127	5.9	370	70.2
S17	131	5.5	420	299.4
S10	128	6.3	102	428

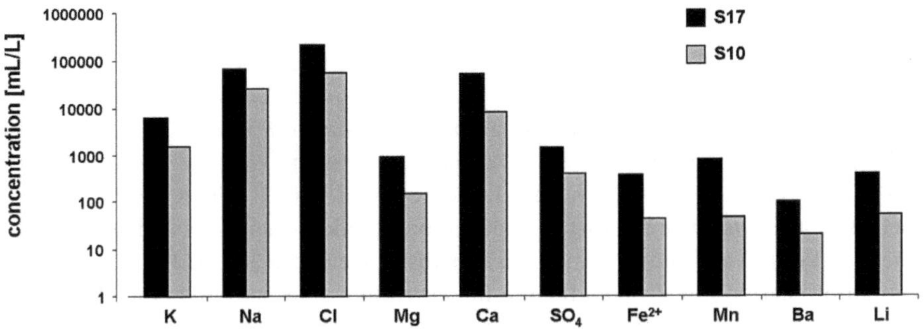

Fig. 5.16 Element- and ion concentrations in reservoir fluids at the two different sites S10 and S17

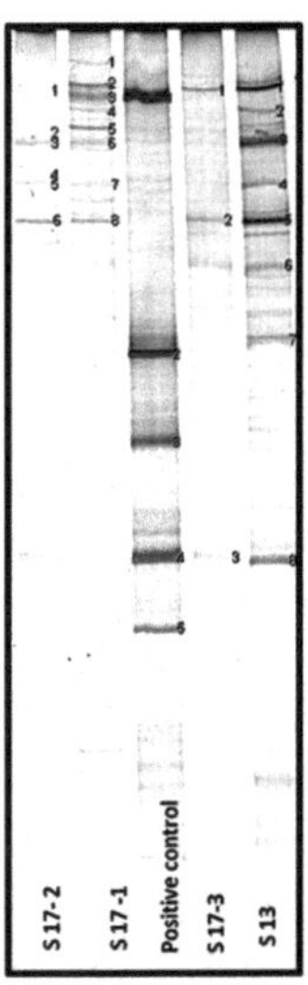

SSCP Fragments	Similarity [%]	Organism	References
S13-1 S17-3-1	100	*Acidovorax sp.*	Indiragandhi, P. et al. Cultivable bacterial strains isolated from two-spotted spider mite Tetranychus urticae population resistance to pyridaben, unpublished
S17-1-2	99	*Comamonas nitrativorans*	Etchebehere, C. et al. Comamonas nitrativorans sp. nov., a novel denitrifier isolated from a denitrifying reactor treating landfill leachate
S17-1-3 S17-3-1 S13-1	99	*Diaphorobacter nitroreducens*	Jroundi, F.F. et al. Culturable carbonatogenic bacteria activated by a consolidating treatment of deteriorated monumental calcarenite stone, unpublished Sun, J. et al. Isolation and characterization of 4-chlorophenol degrading strain from cooking water, unpublished
S17-1-5	96	*Prevotella sp.*	Adolphe, Y. et al. Minimal diversity in grease trap revealed by molecular analysis of bacterial community, unpublished Ueki, A. et al. Prevotella paludivivens sp. nov., a novel strictly anaerobic, Gram-negative, hemicellulose-decomposing bacterium isolated from plant residue and rice roots in irrigated rice-field soil, 2007
S17-1-6 S17-2-3 S13-2 S13-3	99	*Pseudomonas sp.*	Wei, Y. et al. Phylogenetic ananlysis the diversity of Pseudomonas strains isolated from Mingyong glacier, unpublished
S17-1-8 S17-2-6 S17-3-2 S13-5	94	*Ralstonia sp.*	Konstantinidis, K.T. et al. Microbial diversity and resistance to copper in metal-contaminated lake sediment, 2003
S13-3420-8	100	Unidentified bacterium	Boivin-Jahns, V. et al. Bacterial diversity in a deep subsurface clay environment, 1996

Fig. 5.17 The SSCP (*left*) and DGGE (*right*) analyses and sequences affiliation of SSCP/DGGE fragments of the downhole fluid samples from the wells S13 and S17 from the Rotliegend reservoir

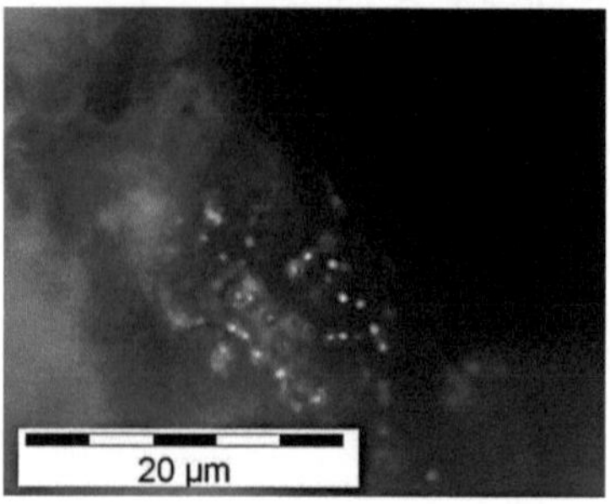

Fig. 5.18 Fluorescent image of the microbial cells sticking to a particle, which are stained with the DNA-dye SYBR-Green. The sample is from reservoir fluids of the well S17

thermophilic communities were also described (Duncan et al. 2009). The genus sphingomonas is characterised by diverse metabolic capability; many members of the genus possess unique abilities in degrading hydrocarbons (White et al. 1996). Some members of sphingomonas are involved in metal corrosion (Arens et al. 1995). The representatives of pantoea sp. have extensive metabolic capabilities under anaerobic conditions, which are capable of growing via dissimilatory reduction of Fe(III), Mn(IV), AQDS, and the toxic metal Cr(VI) (Francis et al. 2000). Species of bacillus sp. are known as extremophilic microorganisms and are capable to reduce Fe (III) and Mn (IV) in the deep subsurface (Boone et al. 1995). By using high amplification rates/cycles for the fingerprinting small amounts of DNA from inhabitants of the reservoir that corresponds to low cell numbers can be detected. However, the high amplification rates increase the likelihood of contamination. Quantification revealed very low cell numbers in the reservoir fluids. With SYBR Green cells in reservoir fluids were detectable and usually attached to particles (Fig. 5.18), but the cell numbers were too low, for exact quantification.

Recent studies showed that upper the temperature limit for microbial life was underestimated (Kashefi 2004; Kashefi et al. 2008). For example, Kashefi and Lovley (2003) were able to cultivate archaea at 121 °C, that were isolated from a 300 °C environment. Other archaeon from deep hydrothermal vent, methanopyrus kandleri, was shown to be able to survive and reproduce at 122 °C (Takai et al. 2008). Furthermore, heterotrophic microorganisms were found in 140 °C hot environments (Ver Eecke et al. 2009). Recent studies revealed that high salinity is one of the important factors supporting DNA stability at very high temperatures (Marguet and Forterre 2001; Daniel et al. 2003), which support the possibility that the observed microorganisms can be active in the extreme habitat although activity methods have to be improved to further investigate this.

With decreasing temperature following the CO_2 injection, higher numbers and activity of microorganisms are expected. This would be especially relevant for the near well area of the injection well and leaking faults of storage sites. Identification of the biocorrosive thermophilic microorganisms could be relevant for the technical progress of the CO_2 storage technique. Recent investigations showed that members of this group were able to rapidly and massively change the permeability of the injectivity in the area most proximal to the well bore (Morozova et al. 2009; Zettlitzer et al. 2010). Thermophilic sulphate reducing bacteria and other thermophilic microorganisms (methanogenic archaea, Fe-reducing bacteria etc.) were discovered to contribute to corrosion of metallic oilfield pipelines (Duncan et al. 2009).

In addition, increasing biomass and mineral precipitation could support reservoir sealing, known as self-sealing process. The results of the long-term incubations of the rock core material with reservoir fluids under in situ reservoir conditions will provide more information about influence of microbial community on creation and dissolution of minerals and the impact on the technical effectiveness of EGR. Since mineralization like other processes in the deep biosphere is a very slow process and the low cell numbers observed, sensitivity of the microbial activity measurements have to be improved to provide information for modelling the long-term behaviour of injected CO_2 and the influence of microorganisms.

References

Aloisi G, Bouloubassi I, Heijs SK, Pancost RD, Pierre C, Sinninghe Damsté JS, Gottschal JC, Forney LJ, Rouchy J-M (2002) CH_4-consuming microorganisms and the formation of carbonate crusts at cold-seeps. Earth Planet Sci Lett 203:195–203

Aragno M, Schlegel HG (1992) The mesophilic hydrogen oxidizing (Knallgas) bacteria. In: Balows HGTpA, Dworkin M, Harder W, Schleifer K-H (eds) The prokaryotes. Springer, New York, pp 3917–3933

Arens P, Tuschewitzki M, Wollmann H, Follner H, Jacob HJ (1995) Indicators for microbiologically induced corrosion of copper pipes in cold-water plumbing system. Zentralbl Hyg Umweltmed 196:444–456

Becker V, Myrttinen A, Blum P, van Geldern R, Barth JAC (2011) Predicting $\delta^{13}C_{DIC}$ dynamics in CCS: a scheme based on a review of inorganic carbon chemistry under elevated pressures and temperatures. Int J Greenhouse Gas Control 5(5):1250–1258

Beech I, Sunner J (2004) Biocorrosion: towards understanding interractions between biofilms and metals. Curr Opin Biotechnol 15:181–186

Bohlen T (2002) Parallel 3-D viscoelastic finite difference seismic modeling. Comput Geosci 28:887–899

Boone DR, Liu Y, Zhao ZJ, Balkwill DL, Drake GR, Stevens TO, Aldrich HC (1995) Bacillus infernus sp. nov., an Fe(III)- and Mn(IV)-reducing anaerobe from the deep terrestrial subsurface. Int J Syst Bacteriol 45(3):441–448

Brown GA, Kennedy B, Meling T (2000) Using fibre-optic distributed temperature measurements to provide real-time reservoir surveillance data on Wytch Farm field horizontal extended-reach wells, SPE 62952. Paper presented at SPE annual technical conference and exhibition, Dallas, 1–4 October

Clark ID, Fritz P (1997) Environmental isotopes in hydrogeology. CRC Press/Lewis, Boca Raton

D'Hondt S, Jorgensen BB, Miller DJ, Batzke A, Blake R, Cragg BA, Cypionka H, Dickens GR, Ferdelman T, Hinrichs KU, Holm NG, Mitterer R, Spivack A, Wang G, Bekins B, Engelen B, Ford K, Gettemy G, Rutherford SD, Sass H, Skilbeck CG, Aiello IW, Guerin G, House CH, Inagaki F, Meister P, Naehr T, Niitsuma S, Parkes RJ, Schippers A, Smith DC, Teske A, Wiegel J, Padilla CN, Acosta JL (2004) Distributions of microbial activities in deep subseafloor sediments. Science 306(5705):2216–2221

Daniel RM, Holden JF, van Eckert R, Truter J, Cowan DA (2003) The stability of biomolecules and the implications for life at high temperatures. In: Wilcock WSD, Cary C, DeLong E, Kelley DS, Baross JA (eds) Subsurface biosphere at mid-ocean ridges. American Geophysical Union Monograph, Washington, DC

Douglas S, Beveridge TJ (1998) Mineral formation by bacteria in natural microbial communities. FEMS Microbiol Ecol 26:79–88

Drever JI (1997) The geochemistry of natural waters: surface and groundwater environments. Prentice Hall, New Jersey

Duan Z, Moller N, Weare JH (1992) An equation of state for the CH_4-CO_2-H_2O system: II. Mixtures from 50 to 1000°C and 0 to 1000 bar. Geochimica Et Cosmochimica Acta 56 (7):2619–2631

Duncan KE, Gieg LM, Parisi VA, Tanner RS, Tringe SG, Bristow J, Suflita JM (2009) Biocorrosive thermophilic microbial communities in Alaskan North Slope oil facilities. Environ Sci Technol 43(20):7977–7984

Francis CA, Obraztsova AY, Tebo BM (2000) Dissimilatory metal reduction by facultative anaerobe Pantoea agglomerans SP1. Appl Environ Microbiol 66:543–548

Gajewski D, Tessmer E (2005) Reverse modelling for seismic event characterization. Geophys J Int 163:276–284

GDF SUEZ (2009) CLEAN DMS data management system, GFZ German Research Centre for Geosciences

Henninges J, Zimmermann G, Büttner G, Schrötter J, Erbas K, Huenges E (2005) Wireline distributed temperature measurements and permanent installations behind casing. In: Proceedings of the world geothermal congress 2005, Antalya [CD-ROM]. Horne R, Okandan E (eds) paper 1021, International Geothermal Association, Reykjavik

Horikoshi K, Grant WD (1998) Extremophiles: microbial life in extreme environments. Wiley-Liss, New York

Hutchens E, Valsami-Jones E, McEldowney S, Gaze W, McLean J (2003) The role of heterotrophic bacteria in feldspar dissolution – an experimental approach. Mineralog Mag 67:1157–1170

Javaherdashti R (2008) Microbiologically influenced corrosion. An engineering insight. Springer, London

Johnson G, Mayer B (2011) Oxygen isotope exchange between H_2O and CO_2 at elevated CO_2 pressures: implications for monitoring of geological CO_2 storage. Appl Geochem 26 (7):1184–1191

Jorgensen BB, Boetius A (2007) Feast and famine–microbial life in the deep-sea bed. Nat Rev Microbiol 5(10):770–781

Kashefi K (2004) Response to Cowan: the upper temperature for life – where do we draw the line? Trends Microbiol 12:60–62

Kashefi K, Lovley DR (2003) Extending the upper temperature limit for life. Science 301 (5635):934–934

Kashefi K, Shelobolina ES, Elliott WC, Lovley DR (2008) Growth of thermophilic and hyperthermophilic Fe(III)-reducing microorganisms on a ferruginous smectite as the sole electron acceptor. Appl Environ Microbiol 74(1):251–258

Kervévan C, Azaroual M, Durst P (2005) Improvement of the calculation accuracy of acid gas solubility in deep reservoir brines: application to the geological storage of CO_2. Oil Gas Sci Technol – Rev IFP 60(2):357–379

Konstantinidis KT, Isaacs N, Fett J, Simpson S, Long DT, Marsh TL (2003) Microbial diversity and resistance to copper in metal-contaminated lake sediment. Microb Ecol 45(2):191–202

Little BJ, Lee JS (2007) Microbiologically influenced corrosion. In: Winston R (ed) Wiley series in corrosion. Wiley, New Jersey

Lüders V, Plessen B, Romer RL, Weise SM, Banks DA, Hoth P, Dulski P, Schettler G (2010) Chemistry and isotopic composition of Rotliegend and Upper Carboniferous formation waters from the North German Basin. Chem Geol 276(3–4):198–208

Marchesi JR, Weightman AJ, Cragg BA, Parkes J, Fry JC (2001) Methanogen and bacterial diversity and distribution in deep gas hydrate sediments from the Cascadia Margin as revealed by 16S rRNA molecular analysis. FEMS Microbiol Ecol 34:221–228

Marguet E, Forterre P (2001) Stability and manipulation of DNA at extreme temperatures. Methods Enzymol 334:205–215

Millero F, Huang F, Graham T, Pierrot D (2007) The dissociation of carbonic acid in NaCl solutions as a function of concentration and temperature. Geochimica Et Cosmochimica Acta 71(1):46–55

Morozova D, Wandrey M, Vieth A, Würdemann H (2009) Microbial community characterisation of the CO_2 reservoir in saline aquifer in Ketzin, Germany. Biospectrum, 214

Morozova D, Wandrey M, Alawi M, Zimmer M, Vieth A, Zettlitzer M, Würdemann H (2010) Monitoring of the microbial community composition in saline aquifers during CO_2 storage by fluorescence in situ hybridisation. Int J Greenhouse Gas Control 4:981–989

Muyzer G, de Waal EC, Uitterlinden AG (1993) Profiling of complex microbial populations by denaturing gradient gel electrophoresis analysis of polymerase chain reaction-amplified genes coding for 16S rRNA. Appl Environ Microbiol 59(3):695–700

Myrttinen A, Becker V, van Geldern R, Barth JAC, Nowak M (2010a) Stable isotope and dissolved inorganic carbon sampling, monitoring and analytical methods in CO_2 injection and EGR projects. In: HouMichael Z, Yoon XH (eds) Underground storage of CO_2 and energy. CRC Press, Jeoungseok, p 384

Myrttinen A, Becker V, van Geldern R, Würdemann H, Morozova D, Taubald H, Blum P, Barth JAC (2010b) Carbon and oxygen isotope indications for CO_2 behaviour after injection: first results from the Ketzin Site (Germany). Int J Greenhouse Gas Control 4(6):7

Orita M, Iwahana H, Kanazawa H, Hayashi K, Sekiya T (1989) Detection of polymorphisms of human DNA by gel electrophoresis as single-strand conformation polymorphisms. Proc Natl Acad Sci USA 86(8):2766–2770

Parkes RJ, Cragg BA, Wellsbury P (2000) Recent studies on bacterial populations and processes in subseafloor sediments: a review. Hydrogeol J 8(1):11–28

Pernthaler A, Pernthaler J, Amann R (2002) Fluorescence in situ hybridization and catalyzed reporter deposition for the identification of marine bacteria. Appl Environ Microbiol 68 (6):3094–3101

Pikuta EV, Hoover RB, Tang J (2007) Microbial extremophiles at the limits of life. Crit Rev Microbiol 33(3):183–209

Rothschild LJ, Mancinelli RL (2001) Life in extreme environments. Nature 409(6823):1092–1101

Rudledge JT (2009) Interpreting reservoir microseismicity detected during CO_2 injection at the Aneth Oil Field. American Geophysical Union, Fall Meeting 2009, U41B-0021

Sahl JW, Schmidt R, Swanner ED, Mandernack KW, Templeton AS, Kieft TL, Smith RL, Sanford WE, Callaghan RL, Mitton JB, Spear JR (2008) Subsurface microbial diversity in deep-granitic-fracture water in Colorado. Appl Environ Microbiol 74(1):143–152

Sakurai S, Ramakrishnan TS, Boyd A, Mueller N, Hovorka S (2005) Monitoring saturation changes for CO2 sequestration: petrophysical support of the Frio brine pilot experiment. In: 46th annual logging symposium transactions, edited, Society of Petrophysicists and Well Log Analysts, New Orleans

Sanchez-Roman M, Rivadeneyra MA, Vasconcelos C, McKenzie JA (2007) Biomineralization of carbonate and phosphate by moderately halophilic bacteria. FEMS Microbiol Ecol 61:273–284

Schippers A, Neretin LN (2006) Quantification of microbial communities in near-surface and deeply buried marine sediments on the Peru continental margin using real-time PCR. Environ Microbiol 8(7):1251–1260

Schwieger F, Tebbe CC (1998) A new approach to utilize PCR-single-strand-conformation polymorphism for 16S rRNA gene-based microbial community analysis. Appl Environ Microbiol 64(12):4870–4876

Setzmann U, Wagner W (1991) A new equation of state and tables of thermodynamic properties for methane covering the range from the melting line to 625 K at pressures up to 1000 MPa. J Phys Chem Ref Data 20(6):1061–1151

Smolen J (1996) Cased hole and production log evaluation. Pennwell, Tulsa

Snieder R, Grêt A, Douma H, Scales J (2002) Coda wave interferometry for estimating nonlinear behavior in seismic velocity. Science 295:2253–2255

Span R, Wagner W (1996) A new equation of state for carbon dioxide covering the fluid region from the triple-point temperature to 1100 K at pressures up to 800 MPa. J Phys Chem Ref Data 25(6):1509–1596

Span R, Lemmon EW, Jacobsen RT, Wagner W, Yokozeki A (2000) A reference equation of state for thermodynamic properties of nitrogene for temperatures from 63.151 to 1000 K and pressures to 2200 MPa. J Phys Chem Ref Data 29(6):1361–1433

Stohr R, Waberski A, Liesack W, Volker H, Wehmeyer U, Thomm M (2001) Hydrogenophilus hirschii sp. nov., a novel thermophilic hydrogen-oxidizing beta-proteobacterium isolated from Yellowstone National Park. Int J Syst Evol Microbiol 51(Pt 2):481–488

Takai K, Nakamura K, Toki T, Tsunogai U, Miyazaki M, Miyazaki J, Hirayama H, Nakagawa S, Nunoura T, Horikoshi K (2008) Cell proliferation at 122 degrees C and isotopically heavy CH_4 production by a hyperthermophilic methanogen under high-pressure cultivation. Proc Natl Acad Sci USA 105(31):10949–10954

Ver Eecke HC, Kelley DS, Holden JF (2009) Abundance of hyperthermophilic autotrophic Fe(III) oxide reducers and heterotrophs in hydrothermal sulphide chimneys of the North-eastern Pacific ocean. Appl Environ Microbiol 75:242–245

Videla HA, Herrera LK (2005) Microbiologically influenced corrosion: looking to the future. Int Microbiol 8:169–180

Vorhies JS, Gaines RR (2009) Microbial dissolution of clay minerals as a source of iron and silica in marine sediments. Nature Geosci 2:221–225

Vu-Hoang D, Henninges J, Hurter S (2009) Monitoring carbon dioxide saturation through wireline logging techniques. EGU Geophysical research abstracts, 11, EGU General Assembly, Vienna, EGU2009-10051

Weinbauer MG, Beckmann C, Hofle MG (1998) Utility of green fluorescent nucleic acid dyes and aluminum oxide membrane filters for rapid epifluorescence enumeration of soil and sediment bacteria. Appl Environ Microbiol 64(12):5000–5003

Welch SA, McPhail DC (2003) Mobility of major and trace elements during biologically mediated weathering of granite. In: Roach IC (ed), Advances in regolith, pp 437–440, CRC for Landscape Environments and Mineral Exploration, PO Box 1130, Bentley, Western Australia 6102 Australia. ISBN 0-7315-4815-9 (CD-ROM)

White DC, Sutton SD, Ringelberg DB (1996) The genus Sphingomonas: physiology and ecology. Curr Opin Biotechnol 7:301–306

Wilhelms A, Larter SR, Head I, Farrimond P, di-Primio R, Zwach C (2001) Biodegradation of oil in uplifted basins prevented by deep-burial sterilization. Nature 411(6841):1034–1037

Xue Z, Ohsumi T (2004) Seismic wave monitoring of CO_2 migration in water-saturated porous sand-stone. Exp Geophys 35:25–32

Zettlitzer M, Moeller F, Morozova D, Lokay P, Würdemann H (2010) Re-establishment of the proper injectivity of the CO_2-injection well Ktzi201 in Ketzin, Germany. Int J Greenhouse Gas Control 4:952–959

Zhou R, Huang L, Rutledge JT, Fehler M, Daley TM, Majer EL (2010) Coda-wave interferometry analysis of time-lapse VSP data for monitoring geological carbon sequestration. Int J Greenhouse Gas Control 4:679–686

Zimmermann RW, Somerton W, King M (1986) Compressibility of porous rocks. J Geophys Res 91:12765–12777

Chapter 6
Environmental and Process Monitoring

Dirk Schäfer, Said Al Hagrey, Esben Auken, Andreas Bahr, Matthias Beyer, Andreas Dahmke, Ingolf Dumke, Nikolaj Foged, Markus Furche, Michael Gräber, Jochen Großmann, Max Helkjaer, Ralf Köber, Jürgen Poggenburg, Gert Naue, Stefan Schlömer, Christian Seeger, Lars Tischer, Angelika Vidal, Carla Wiegers and Christian Wöhrl

Abstract For enhanced gas recovery (EGR) using CO_2 as well as for CO_2 storage in depleted gas fields it needs to be shown that injection and storage is save and neither population nor environment is exposed to risks during operation or afterwards. This requires the development and application of methods to monitor groundwater, vadose zone and atmosphere. Therefore, extensive investigations of the near-surface aquifers were performed to characterize the geological structure and the geochemical and hydraulic conditions as part of a baseline-monitoring and to specify input parameters for model simulations. If CO_2 leakage should occur and CO_2 migrates upwards from the storage complex, shallow freshwater aquifers are the first protected good that might be affected. Based on the model simulations, parameters that would be affected by leakages were specified and parameter changes as well as spatial extension of the expected changes quantified. A comparison of the model results with measured natural variabilities show that especially pH and TIC (total inorganic carbon), but under certain conditions also electric

D. Schäfer (✉) • S. Al Hagrey • A. Dahmke • R. Köber • C. Wiegers
Christian-Albrechts-Universität of Kiel Ludewig-Meyn-Str. 10, 24118 Kiel, Germany
e-mail: ds@gpi.uni-kiel.de

E. Auken • N. Foged
Department of Earth Sciences, University of Aarhus, Høegh-Guldbergs Gade 2, Århus C 8000, Denmark

A. Bahr • I. Dumke • M. Furche • J. Poggenburg • S. Schlömer • C. Seeger • A. Vidal • C. Wöhrl
Federal Institute for Geosciences and Natural Resources (BGR), Stilleweg 2, 30655 Hannover, Germany

M. Beyer • J. Großmann • G. Naue • L. Tischer
GICON – Großmann Ingenieur Consult GmbH, Tiergartenstraße 48, 01219 Dresden, Germany

M. Gräber
Geoserve – Angewandte Geophysik, Lützowstraße 3a, 24105 Kiel, Germany

M. Helkjaer
SkyTEM Surveys ApS, Tofteledet 18, Beder DK - 8330, Denmark

M. Kühn and U. Münch (eds.), *CLEAN: CO_2 Large-Scale Enhanced Gas Recovery in the Altmark Natural Gas Field*, Advanced Technologies in Earth Sciences, DOI 10.1007/978-3-642-31677-7_6, © Springer-Verlag Berlin Heidelberg 2013

conductivity and aqueous calcium concentration (Ca) are most suited parameters for the detection of CO_2 leakages based on observation wells in shallow aquifers. It was an important result that the temporal fluctuations of groundwater composition are generally small but spatial variations are large.

The simulation results demonstrate that CO_2 gas and dissolved inorganic carbon should be expected and measured in the upper part of a confined aquifer. They also show that CO_2 gas will follow the steepest gradient of the overlying aquitard and accumulate in anticlinal structures, while the dissolved inorganic carbon migrates in groundwater flow direction.

Since the area that can be monitored by observation wells and also the extension of leakage plumes are limited, it is advised to additionally use further, e.g. geophysical monitoring techniques, which are able to cover wider areas. The tested methods ERT (electrical resistivity tomography) and especially SkyTEM (helicopter borne electromagnetic survey) can provide quite comprehensive information about the aquifer down to a depth of 300 m. This is to image the geological structure, which would have the strongest influence on the distribution of rising CO_2 gas and is therefore important for a site-specific monitoring concept. Repeated aero-electromagnetical measurements are so far the only method that could be used for an area-wide leakage monitoring on the relevant scale. All current laws and regulations demand an "adequate monitoring", but it will be an iterative discussion in the future between operators, public, authorities and scientists to define what should be measured, how often and where.

In case of a large CO_2 leakage rate the gas penetrates the shallow aquifer structures and migrates into the vadose zone. Here the CO_2 can be measured before it discharges into the atmosphere. The CO_2 concentrations in the soil are temporarily variable due to microbial degradation of organic carbon. Soil gas monitoring can be a suitable and sensitive additional method to detect non-natural CO_2 concentrations in the vadose zone. The preconditions are the possibility to measure the CO_2 concentration significantly below microbiologically active soil zones but above varying ground or tail water levels. Both constraints have to be tested by preliminary site investigations. Because of strong local and seasonal variations individual baselines for at least 2 years are required for interpretation of the CO_2 concentration measurements.

CO_2 might leak through the geological formation into the atmosphere, but it can also be released due to accidents in the surface facilities. Only very few investigations exist how CO_2 spreads in the atmosphere and where hazardous concentrations are reached. Where CO_2 is decanted or handled the potential hazards of CO_2 release at the surface or around the facilities need to be considered. Calculations emphasize that the hazardous areas resulting from unwanted releases of CO_2 cannot simply be estimated. Specific hazards must be determined in an event- and location-specific manner. Nevertheless, in that way site-specific prognoses and thus the introduction of specific monitoring and intervention measures are possible. The simulations demonstrate that simple recommendations can improve safety, like turning the exhaust of the safety valves vertically upwards to enhance mixing in the atmosphere.

It is outlined here that many different aspects with regard to the monitoring of atmosphere, vadose zone and groundwater need to be taken into account with regard to EGR and CO_2 storage. In a future step a coherent monitoring concept needs to be defined now for a whole site on the demonstration scale. An intensive discussion between legal authorities and scientist will be required to harmonize what is technically feasible and what is socially and legally expected.

6.1 Development of Monitoring Concepts for Geologically Stored CO_2

For the geological storage of CO_2 like in EGR or CCS projects it has to be proven by a set of suited observations that the storage is save and neither population nor environment is exposed to risks during operation and after site closure. Although the operation of a storage site will only be permitted if different risks that theoretically can arise have been shown to be low within the site characterization, a site-specific monitoring concept has to be developed to ensure that the project execution is ongoing as intended and that no protected natural recourses are at risk. Such a monitoring concept, which is called monitoring plan in the EU CCS Directive 2009/31/EC, has to comprise all monitoring compartments from the storage formation to the atmosphere and has to cover the total project duration from the pre-injection phase, when baseline conditions are measured as a reference state, to the post-closure phase, after the injection is finished and monitoring is still going on to guarantee save storage. A detailed report with recommendations for the development of such monitoring concepts focusing on protected natural recourses was prepared during the CLEAN project (Köber and Dahmke 2011). Extracts from the report are given in the following.

The extent of a monitoring concept is primarily dependent on the imposed demands, which were defined in the EU CCS Directive, which all member states had to transfer to national law. The targets of monitoring given therein are the detection of CO_2 migration, CO_2 leakage, significant irregularities, and significant adverse effects for the surrounding environment, including in particular drinking water, human population, and the surrounding biosphere. Updating the assessment of the safety and integrity of the storage complex in the short and long term, including the assessment of whether the stored CO_2 will be completely and permanently contained is also demanded. The modelled behaviour of CO_2 and the formation water at the storage site has to be compared to actual measurements. In case of application of corrective measures, their effectiveness has to be assessed. The directive does not prescribe the parameters to be monitored, the methods to be used and the location and frequency of measurement. These aspects have to be specified for each monitoring phase in a monitoring plan. The only concrete aspects that are demanded in any case are continuous or intermittent monitoring of: (1) fugitive emissions of CO_2 at the injection facility, (2) CO_2 volumetric flow at injection wellheads, (3) CO_2 pressure and temperature at injection wellheads (to determine mass flow), (4) chemical analysis of the injected material, and (5) reservoir temperature and pressure (to determine CO_2 phase behaviour and state). These obligatory aspects were supplemented in the German CCS draft law by the monitoring of the chemical and physical quality of the ground water.

Beside these legal requirements the following important aspects should additionally be accounted for in monitoring concepts, as pointed out in a report for the German Federal Environment Agency (UBA) on the safety and environmental compatibility of CO_2 storage (Großmann et al. 2011): (1) microseismics induced by CO_2 injection, (2) uplift or subsidence of land surface, (3) pressure propagation also beyond the storage formation, (4) leakage of other gases than CO_2 (e.g. methane), (5) integrity of abandoned wells, (6) changes in the salt-fresh-water-boundary, (7) conflicts with respect to the utilisation of other geological resources. Especially for EGR sites leakage of other gases like methane and the integrity of abandoned wells are of particular importance.

The UBA CCS report (Großmann et al. 2011) seems to be so far the most systematic approach for the evaluation of the safety and the environmental compatibility of CO_2 storage in Germany. It critically examines with definitions given in the EU CCS Directive and emphasises the need of conceptual clarification. The report suggests evaluating the technology in accordance with the German Major Accidents Ordinance in five steps: (1) Defining the terminology for system explanation, (2) characterization of storage related processes and (3) its environmental impacts, (4) description of model simulations and methods for site characterization and monitoring, and (5) risk management including a description of intervention measures. While all supposed processes for the specified service of storage have to be monitored, a prioritisation of processes, that only take place in the case of unexpected incidents, is recommended. This prioritisation should be classified in the three categories: (1) reasonable non-excludable hazard sources, (2) reasonably excludable hazard sources, and (3) exceptional hazard sources which defy all predictability.

The most important issues in the development of monitoring concepts are the decisions about the methods which can be used to meet the individual monitoring objectives and the dimensioning of the respective method application. The CCS Directive does only indicate that the choice of monitoring technology has to be based on the best practice available at the time of design, and the decisions have to be justified. During the last 3 years several studies like the NETL report (NETL 2009) or the UBA CCS report (Großmann et al. 2011) mentioned above focused on the objectives of monitoring and also gave examples of monitoring technologies that could be used to achieve these objectives. Such recommendations are quite helpful at the present state of development, but lack in empirical reference data for the proposed methods. This would be important for the dimensioning of individual method applications and the overall development of monitoring concepts. In terms of monitoring optimization and security of CCS or EGR applications and regarding the current state of knowledge, more quantitative information about monitoring methods is required. One focus within CLEAN was therefore as a first step to create a comprehensive compilation of more than 60 existing monitoring methods with method descriptions, literature references, detection limits, costs, development status as well as pros and cons. Such characteristics were compiled and made available for a range of methods in the form of method profiles.

Method sensitivity is one of the key aspects to judge, whether a method is suited to meet a defined goal or not. For most methods the detection limit with regard to the primary measured parameter is known, but this parameter is in most cases not directly the relevant monitoring parameter and therefore no direct measure for sensitivity. It is therefore frequently necessary to transfer the measured parameter into the value of interest. Sensitivity criteria for instance for leakage detection should be indicated in a unit like detectable CO_2 mass per volume rock or leakage rate, but possible methods for leakage detection like seismics, geoelectrics, or ground water analyses do not measure these values. Furthermore it is not possible to convert the measured values into the target parameters by straightforward calculations. Generally it is necessary to generate scenarios, define boundary conditions, make assumptions, and consider different physical or chemical processes to obtain the relevant information. Because of the complexity of these coherences, such calculations have generally to be done by numerical simulation. For most technologies or monitoring compartments such simulations have not been performed so far, but are of particular interest for a more quantitative evaluation of method capabilities. In the CLEAN project these kinds of scenario simulations were done to evaluate the available methods for CO_2 leakage detection in near-surface ground water aquifers (see below). The multi-phase-flow-scenarios considered variabilities in the leakage rate, the composition of rock and ground water, and ground water flow velocities. On the basis of site-specific simulations it was possible to (1) determine the most suitable monitoring parameters, (2) distinguish between natural and CO_2 leakage induced changes, (3) suggest the most suitable monitoring positions and (4) specify the distances between monitoring wells in dependence of the leakage rate for which its impacts should be detectable at minimum.

So far there are no examples of already developed and available monitoring plans and no sublegal regulations exist which could be taken as a guideline for the development of monitoring concepts. It is not foreseeable at the present state what efforts will be demanded by the authorities to fulfil the monitoring objectives from the EU CCS Directive and if the additional objectives mentioned above (UBA CCS report, Großmann et al. 2011), which are still under discussion, or further objectives (for instance as from NETL 2009) have to be included into a monitoring concept. To provide project managers with a broad information basis that can be used as a flexible support in the development of site-specific monitoring concepts, a comprehensive matrix was prepared containing on the one hand all the different objectives from the EU CCS Directive as well as from the UBA and NETL reports, and about 60 monitoring methods on the other hand. For each of the partially overlapping 30 objectives it is indicated in the matrix for all 60 methods whether the method can contribute to the accomplishment of the defined goals or not. Since firstly the establishment of some methods is still under progress, secondly conditions at each site will be different, and thirdly responsible authorities will impose specific requirements, the matrix was kept as general as possible at this state and does not prioritize particular methods. The matrix was developed on basis of a literature study and experiences made during the CLEAN project.

Because some objectives especially from the EU CCS Directive are kept very general and unspecific, and since a relatively large number of monitoring methods are available for individual objectives in the matrix, recommendations for preferred methods were prepared in a next step. Many of the monitoring methods are capable to contribute to multiple objectives, but their application is mostly focused on a specific compartment like for instance reservoir, aquifers, or atmosphere. It is therefore more useful and clearly to refer recommendations to monitoring compartments than to the wide variety of individual monitoring objectives. The monitoring compartments used for method recommendations were classified into the compartments storage formation, well bores, retention formation, transfer formation, and protected natural resource formation. This classification, which is based on the German Major Accidents Ordinance, was developed across projects and is described in the UBA CCS report (Großmann et al. 2011) in more detail.

Especially the lateral extension of these monitoring compartments influences the effort and extent of the required monitoring measures. The lateral extensions are dependent on the expected system changes like for instance CO_2 migration or leakage, pressure propagation, rock deformation, etc. and have to be derived from numerical simulations for the respective formations. Such simulations have to be performed firstly for the specified service, but secondly also for the case of unexpected incidents, even if a CO_2 leakage from an EGR reservoir like the Altensalzwedel complex of the CLEAN site is extremely unlikely. The simulation results and determined extension of a formation below have to be known and considered for the simulation of an upper or enclosing formation. This means for example that the lateral extension of the protected natural resource formation can only be determined on a well-founded basis, if simulation results for the underlying storage formation, retention formation, and transfer formation already exist. Whereas model predictions for the specified service in the storage formation are generally demanded and performed during the planning of CCS or EGR projects, further research and model development is still necessary for process predictions in the retention and transfer formations.

Since there is still a lack of information from model simulations and field experiments about method sensitivities, which could be used to optimize the design of monitoring measures in terms of minimizing expenditure, measurement configurations can so far only be estimated from experiences gained from different fields of application. For the optimization of CCS or EGR monitoring strategies it is necessary to be aware of (1) the detailed geological structure of the different monitoring formations, (2) the interaction of the geological structures with the induced processes and (3) method capabilities. Deeper knowledge about process interactions and their measurability can only be gained from further pilot projects where CO_2 will be injected, as originally intended in the CLEAN project, and by controlled incident field experiments like the ZERT experiment (Spangler et al. 2009, 2010) or the CO_2 injection test in Wittstock (Peter et al. 2011). Furthermore such field tests are crucial for the validation of promising monitoring technologies with regard to their suitability in the field of CCS and EGR, and for the reliable development of monitoring concepts.

6.2 Baseline Monitoring and Simulated Effects of CO_2 Leakages in Near-Surface Aquifers

In the unexpected case of CO_2 influx into near-surface freshwater aquifers, low pH, mineral dissolution, release of inorganic contaminants or increasing salt content could be undesirable impacts. To ensure intactness of the protected natural resource groundwater, legislative directives for CCS or EGR therefore demand the development and implementation of appropriate site-specific groundwater monitoring concepts especially for near-surface aquifers. The development of such concepts requires knowledge about the geological and hydrogeological situation, about the physical and geochemical behaviour of CO_2 under the given conditions and about the capabilities of available monitoring methods. All three aspects therefore were addressed in this subproject to provide necessary information for the development of a monitoring plan (Köber et al. 2011). To decide whether a monitoring method is suited to fulfil requirements or not, to evaluate pros and cons of different monitoring techniques and to conceptualize the application of a chosen monitoring tool the following have to be known:

1. Parameters that could be influenced by CO_2 or saline water intrusions.
2. Intensity of parameter changes expected due to CO_2 or saline water intrusions.
3. Spatial extension of changes expected due to CO_2 or saline water intrusions.
4. Ranges of natural changes for parameters that could be influenced by CO_2 or saline water intrusions (data noise).
5. Method sensitivities (detection limits).
6. Spatial extension of method-specific detection zones.
7. Method-specific assessment of the economic proportionality.

This approach focuses on CO_2 and saline water intrusions, because these two risks are most relevant for near-surface aquifers. For different monitoring compartments (e.g. reservoir, cap rock, atmosphere) or different risks (e.g. seismicity, land deformation, pressure propagation, etc.) comparable evaluation approaches should be formulated. For aspects 1–3 it is necessary to simulate close to reality leakage scenarios using numerical simulation codes and evaluate observed system changes. Boundary conditions and input parameters for such model simulations have to be based on site-specific field and laboratory investigations. Field investigations are also necessary to determine ranges of natural parameter changes.

Method sensitivities should be regarded in terms of how large the mass of leaking CO_2 or saline water has to be, in order to be detectable with a specific method. Method sensitivities are in this sense a priori known only for some methods, e.g. if the measured parameter can be considered as a direct indicator for leakage (e.g. salt content for saline water intrusion). But since most measurements are influenced by a complex set of physical and chemical processes, model calculations should be used to evaluate method sensitivities for different boundary conditions and leakage scenarios.

To develop these necessary assessment criteria, different kinds of site investigations, laboratory analyses and numerical simulations were performed for the shallow aquifers for the Altensalzwedel site of the CLEAN project. Site investigations were based on mature standard technologies, but were gradually supplemented also by innovative methods which still have to prove to be beneficial for CCS or EGR monitoring.

6.2.1 Geologic Structural Model

To model the CO_2 spreading in the upper quaternary layers, a geologic structural model of the investigation area had to be developed. The investigation area is situated between the terminal moraine of the Saalian Glacial Period in the south and the Weichselian Sequence in the north. The area is dominated by plateaus (e.g. Jeetze-Dumme-Plate) and low-lying areas (Jeetze-Purnitz). The plateaus consist mainly of till and melt water sand of the Saalian Glacial Period. They are surrounded by melt water sand-filled erosion gullies. In the low-lying areas, shallow silty Holocene facies are found on top of the sand and gravel deposits. Below these facies, limnic formations of the Eem Interglacial as well as fluvial sands and till of the Saalian Glacial Period are found.

The geologic information for the structural model was available from the quaternary lithofacies maps (LKQ), hydrogeologic maps (HK 50) and boring logs of bore holes from several public and private sources. Two drill cores, as well as information from the drilling progress of 12 new monitoring wells (B1–B12) with an average depth of 50 m, provided further geologic data for the investigation area. The structural model is based on the regional geological structuring. The structural model was developed by executing the following steps:

- Derivation of the extension of geologic units.
- Classification of the observed lithology.
- Derivation of the hydrogeologic unit layer geometry.
- Adaption of the extension of hydrogeologic units.
- Derivation of the layer thickness.
- Set-up of the three-dimensional structural model.

To set up the model, the software package GMS 6.5 (Groundwater Modelling System) from Brigham Young University in Utah, USA, was used. The model domain covers an area of approximately 10 km^2 and is bounded by the Purnitz and Jeetze Rivers to the Northwest and stretches 4 km to the Southeast (Fig. 6.1). The total thickness of the model ranges between 30 and 50 m. Four relevant layers have been identified, as shown in Table 6.1 and Fig. 6.2.

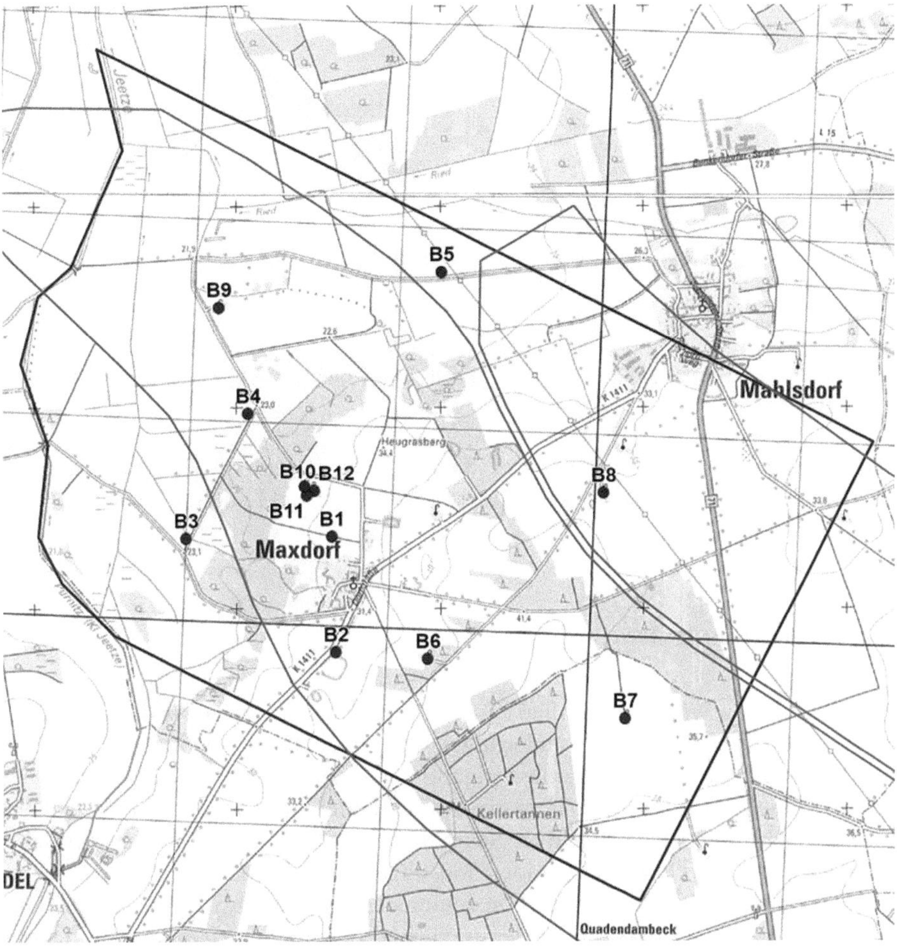

Fig. 6.1 Map of the investigation area with groundwater monitoring wells (B1–B12) and geoelectric surface profiles (S1–S5)

Table 6.1 Layers of the structural model

Layer	Hydrogeologic unit	Aquifer/aquitard
L1	Holocene sediment, till of Saalian II	Aquitard
L2	Weichselian Saalian Glacial Period I/II	Aquifer
L3	Till of Saalian I	Aquitard
L4	Saalian Glacial Period I, Elsterian Glacial Period	Aquifer

On the plateau to the southeast, melt water sand of the Saalian Glacial Period I/II represents the main aquifer beneath a 10–20 m tall till layer of the second Saalian Glacial Period. To the Northwest, the upper aquifer consists of sandy sediments of the Weichselian and the Saalian Glacial Period I/II. It is partly overlaid by silty Holocene sediments. The upper aquifer (Layer 2) is mostly separated from the

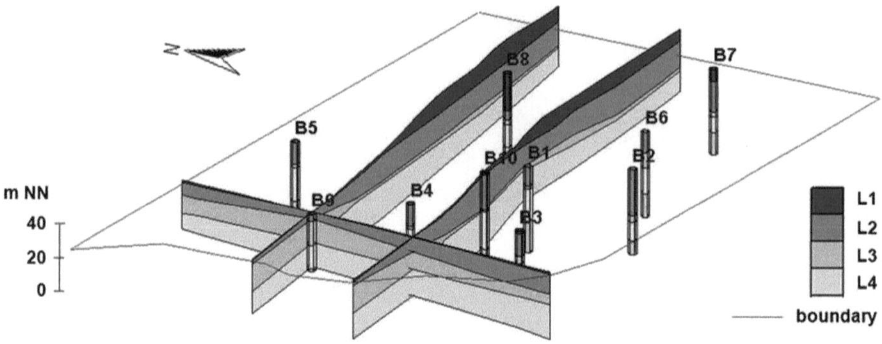

Fig. 6.2 Cross sections of the structural model and geological profiles at the groundwater monitoring wells

lower aquifer by the till of the first Saalian Glacial Period. The lower aquifer consists of fluvial sediments of the Saalian Glacial Period I and the latest Elsterian Glacial Period.

6.2.2 Sediment Analyses

Twelve (12) new wells were drilled to monitor baseline conditions in the uppermost 50 m of the freshwater aquifers at locations which were thought to be representative for the conditions found in the investigation area using sonic drilling technology (Fig. 6.1). Sediment for the characterization of model input parameters was gained from core drillings at B8 and B11. Samples were taken from different depth and were analysed for elemental composition (digestion/ICP-MS), organic (OC) and inorganic carbon (IC) content (carbon coulometer), mineralogy (X-ray diffraction, XRD), hydraulic permeability measurements (sieving/elutriation), and cation exchange capacity.

The elemental compositions of the boulder clay and sandy layers reflect the higher mineralogical diversity in the boulder clay layers compared to the quartz dominated sand layers (Table 6.2). Other mineral phases than quartz could not be detected by XRD measurements from the sandy layers because of the method's detection limit and the low concentrations which could be expected from the elemental composition. XRD results from the boulder clay also proved the existence of calcite besides the main component quartz. Since the mineralogical composition also of minor components is very important for model predictions of groundwater composition changes as a result of CO_2 intrusions, concentrations of typical mineral phases were calculated from the elemental compositions (Table 6.2). These calculations indicate that up to 5 % orthoclase, 2.5 % albite, 1.5 % iron hydroxide, 2.5 % anorthite and 0.4 % calcite can be present in the sand layers, where CO_2 probably would accumulate and react with groundwater and these mineral phases. Results from numerical simulations of such reactions are presented in the modelling chapter.

Table 6.2 Hydraulic conductivities (k), analysed elemental composition and calculated mineral composition of sediment samples from different depths of the drilling cores B8 and B11. Elements in *brackets* indicate from which element the mineral concentration was calculated

Position	Depth	Geology[a]	k-value	IC	OC	Ca	Fe	K	Mg	Mn	Na	Orthoclase $KAlSi_3O_8$ (K)	Albite $NaAlSi_3O_8$ (Na)	Iron hydroxide $Fe(OH)_3$ (Fe)	Anorthite $CaAl_2Si_2O_8$ (Ca)	Calcite $CaCO_3$ (Ca)	Calcite $CaCO_3$ (C)
	[mbs]		[m/s]	[g/kg]	[g/kg]	[g/kg]	[g/kg]	[g/kg]	[g/kg]	[g/kg]	[g/kg]	[g/kg]	[g/kg]	[g/kg]	[g/kg]	[g/kg]	[g/kg]
B8	3–4	bc	3,4E-07	5.33	0.74	29.40	18.41	15.52	4.05	0.26	4.47	110.4	51.0	35.2	203.9	73.4	44.5
B8	5–6	bc	6,9E-07	16.86	1.33	70.38	14.71	11.98	5.14	0.33	4.07	85.2	46.4	28.2	488.1	175.7	140.6
B8	10–11	bc	*1,0E-04*	5.99	1.01	4.22	4.36	5.87	0.77	0.08	1.96	41.7	22.3	8.3	29.3	10.5	50.0
B8	11–12	s	1,3E-04	0.51	0.28	1.79	2.89	4.92	0.48	0.06	1.42	35.0	16.2	5.5	12.4	4.5	4.3
B8	16–17	s	8,3E-05	0.02	0.10	0.51	1.62	4.04	0.25	0.03	1.43	28.7	16.3	3.1	3.5	1.3	0.15
B8	21–22	s	6,4E-05	0.02	0.43	0.53	1.67	4.24	0.20	0.03	1.56	30.2	17.8	3.2	3.6	1.3	0.13
B11	2–4	s	5,8E-05	0.01	1.58	1.47	7.95	6.83	0.59	0.10	2.25	48.6	25.6	15.2	10.2	3.7	0.04
B11	4–6	bc	1,4E-06	0.18	4.72	4.19	14.99	8.46	1.95	0.13	2.25	60.2	25.6	28.7	29.0	10.5	1.51
B11	8–10	s	5,4E-05	0.51	3.66	3.64	7.59	7.39	0.82	0.10	2.07	52.6	23.6	14.5	25.3	9.1	4.28
B11	12–14	s	6,5E-04	0.01	10.40	1.20	3.04	3.13	0.23	0.02	0.91	22.3	10.3	5.8	8.3	3.0	0.06
B11	19–20	s	6,1E-04	0.01	1.42	0.91	3.17	4.29	0.25	0.06	1.65	30.5	18.8	6.1	6.3	2.3	0.09
B11	25–26	s	6,1E-05	0.01	0.91	0.63	2.88	3.28	0.17	0.06	1.25	23.3	14.2	5.5	4.4	1.6	0.06
B11	33–34	s	2,2E-04	0.01	0.35	0.00	1.41	2.60	0.10	0.01	0.17	18.5	1.9	2.7	0.0	0.0	0.06
B11	39–40	s	1,0E-04	0.01	9.72	0.43	1.97	4.14	0.09	0.01	0.31	29.4	3.6	3.8	3.0	1.1	0.07
B11	45–46	s	6,4E-05	0.01	10.32	0.75	3.64	5.03	0.34	0.03	0.33	35.8	3.7	7.0	5.2	1.9	0.07
		bc (min)	3,4E-07	0.18	0.74	4.19	4.36	5.87	0.77	0.08	1.96	41.73	22.29	8.34	29.05	10.46	1.51
		bc (max)	1,4E-06	16.86	4.72	70.38	18.41	15.52	5.14	0.33	4.47	110.41	50.96	35.24	488.06	175.68	140.60
		bc (mean)	8,1E-07	7.09	1.95	27.05	13.12	10.46	2.98	0.20	3.19	74.39	36.31	25.10	187.57	67.51	59.14
		s (min)	5,4E-05	0.01	0.10	0.00	1.41	2.60	0.09	0.01	0.17	18.49	1.91	2.70	0.00	0.00	0.04
		s (max)	6,5E-04	0.51	10.40	3.64	7.95	7.39	0.82	0.10	2.25	52.56	25.64	15.22	25.25	9.09	4.30
		s nmean)	2,0E-04	0.11	3.76	1.04	2.99	4.31	0.29	0.04	1.11	30.63	12.64	5.72	7.20	2.59	0.93

[a]*bc* boulder clay, *s* sand

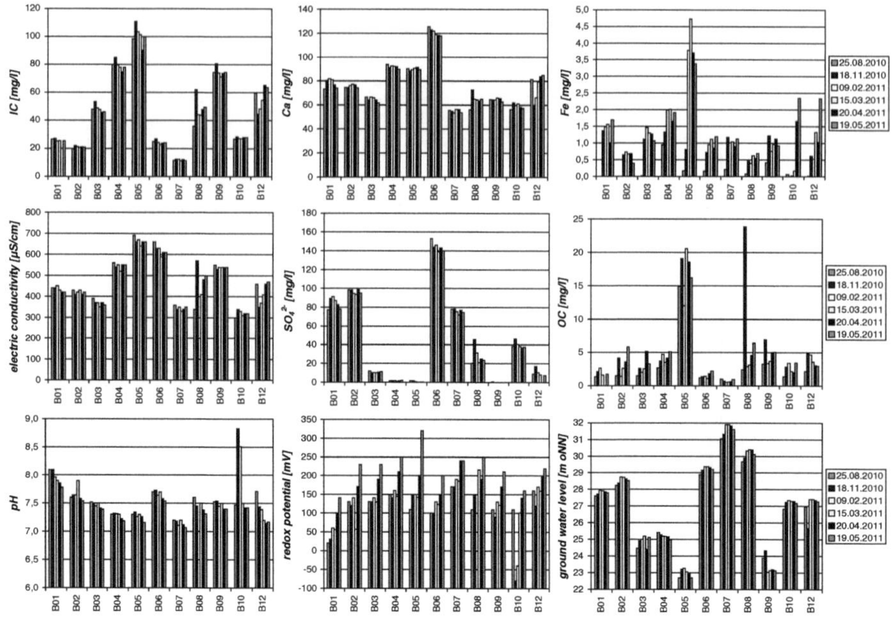

Fig. 6.3 Main parameters from six groundwater sampling campaigns

The calculation of hydraulic conductivity (k) values from sieving and elutriation results yields $3.4 \cdot 10^{-7}$ to $1.4 \cdot 10^{-6}$ m/s for the boulder clay and $5.4 \cdot 10^{-5}$ to $6.5 \cdot 10^{-4}$ m/s for the sand layers, which are typical values for such geological sediments. These data were used for the calculation of flow velocities using the developed flow model and for the investigation of the CO_2 impact under consideration of groundwater flow.

6.2.3 Groundwater Sampling

Groundwater was sampled at the newly drilled groundwater monitoring wells six times during the course of 1 year to investigate temporal and spatial variabilities of natural changes in groundwater composition that were representative for the investigation area. Results indicate that temporal variations at individual monitoring wells are in general much smaller than differences between the wells (Fig. 6.3). Only dissolved iron concentrations showed significant temporal changes which may be related to variable groundwater level and redox processes. It can be concluded that individual monitoring wells cannot be regarded as representative for larger parts of the monitoring area, but that the geochemical baseline conditions in the near-surface aquifer are favourable to detect intrusions of CO_2 or saline water. This was also verified by geochemical modelling (see below).

Isotope measurements that can be used for the characterization of different CO_2 sources delivered $\delta^{13}C$ values between -10 and -16 in the groundwater samples. Since the $\delta^{13}C$ value of injected CO_2 ranges about -28, such isotope measurements could help to distinguish between leaking CO_2 from the reservoir and pre-existing natural CO_2.

Newly developed CO_2 sensors for permanent measurement in groundwater using a gas-selective membrane were also tested at two wells. The measured CO_2 concentrations were quite stable and showed much less changes than CO_2 measurements in the vadose zone. Since the sensors could be equipped with online data transmission and because small changes can be observed in real time, application of such sensors can be very useful for the monitoring of selected areas which are subject of increased risk.

The groundwater flow model, developed using the gained geological information and the measured groundwater levels, indicated flow velocities between 0.1 and 0.4 m/day. This velocity range was also considered in the design of leakage scenario simulations.

6.2.4 Modelling

The geochemical and physical behaviour of CO_2 in near-surface aquifers was investigated using geochemical equilibrium modelling (PhreeqC, Parkhurst and Appelo 1999) and multi-phase-flow modelling (TOUGHREACT EOS 2, Xu et al. 2006). To assess which groundwater monitoring parameters could show measurable changes as a result of CO_2 intrusions, equilibrium concentrations were calculated using 0D batch reactor simulations with present CO_2 gas and different mineral phases. The chosen mineral phases can be assumed to exist in the investigated aquifers due to the sediment analyses. The scenario considering only equilibrium to CO_2 represents reactions of CO_2 with prevailing groundwater in relatively unreactive sediments like quartz sands.

The results showed that pH, electric conductivity, Ca, and TIC (total inorganic carbon) are suited parameters for CO_2 detection under the prevailing site-specific conditions (Fig. 6.4). Whereas pH and TIC show significant changes independently of the mineralogical background, results for Ca and electric conductivity depend strongly on the mineralogical composition. Changes in Ca and electric conductivity would be observed best, if calcite is present.

These results show that changes caused by leaking CO_2 in groundwater geochemistry would be much larger than the natural variabilities which were observed at individual monitoring wells within the course of 1 year and also than the differences between the monitoring wells. This means that it can be evaluated with high probability, if groundwater sampled at monitoring wells was in contact with CO_2 gas in shallow aquifers. The extension of CO_2 gas plumes including the influences of leakage rate, groundwater flow velocity, dissolution and depth thereon was investigated in multi-phase-flow scenarios to assess the probability of capturing a CO_2 plume by monitoring wells.

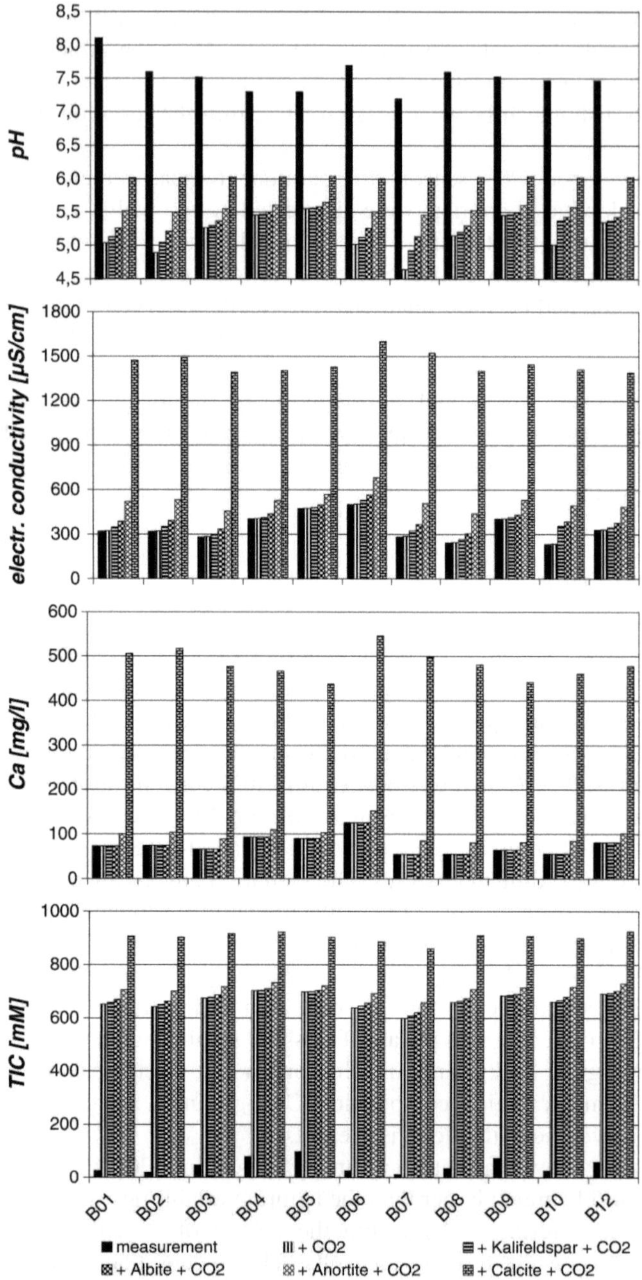

Fig. 6.4 Equilibrium states for most important monitoring parameters calculated by PhreeqC in comparison to the measured values from the first monitoring campaign. Analyzed groundwater compositions were brought into equilibrium only with CO_2 (+CO_2) or with CO_2 and different mineral phases (e.g. +Kalifeldspar+CO_2)

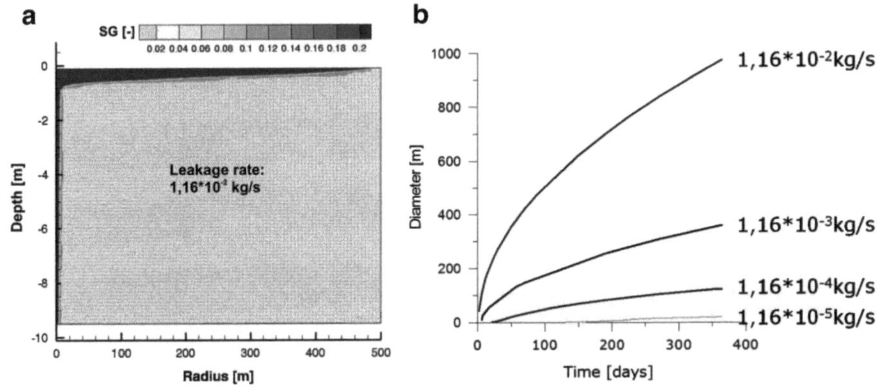

Fig. 6.5 Radial symmetric spreading of CO_2 gas phase after 1 year for a high leakage rate (**a**) and diameters of radial symmetric CO_2 plumes in dependence of leakage rate and time (**b**)

In an unconfined aquifer CO_2 gas migrates upwards from the CO_2 leakage spill to the top of the aquifer and degases into the overlying compartments (i.e. vadose zone). Model simulations showed only very small affected aquifer areas for such scenarios. In a confined aquifer the CO_2 moves upwards until it reaches a lower permeable layer, accumulates and migrates horizontally under the layer and builds a CO_2 pool. Since the detectability of leaking CO_2 in unconfined aquifers is extremely low, groundwater monitoring should focus on confined aquifer sections where CO_2 can spread under clayey or marly aquitards and hereby facilitates its detection. To show the influence of the leakage rate on such spreading CO_2 phase bodies and to quantify the extension of the resulting plumes, 2D radially symmetric models were used with leakage rates from $1.16 \cdot 10^{-2}$ to $1.16 \cdot 10^{-5}$ kg/s.

Figure 6.5 shows the temporal development of CO_2 plume diameters over the leakage time of 1 year. Under such idealized homogenous, planar conditions the resulting diameters would be about 100 m for a leakage rate of $1.16 \cdot 10^{-4}$ kg/s and up to 1,000 m for a leakage rate of $1.16 \cdot 10^{-5}$ kg/s. This indicates that an area-wide observation using monitoring-wells would only be realistic for the highest considered leakage rates. The application of monitoring wells should therefore be focused on areas with increased risk like faults or abandoned wells or on the characterisation of already detected leaks. The results further showed very low thicknesses in the range of decimetres for the spreading plume areas. Monitoring and well filter sections should therefore focus on this area to detect such plumes.

The influence of groundwater flow pore velocities on the extent of the CO_2 gas phase body and the dissolved CO_2 in groundwater was investigated in 3D simulations (Fig. 6.6). The leakage rate was $1.16 \cdot 10^{-5}$ kg/s and pore flow velocities varied in the range of 0.0 to 0.5 m/day according to field measurements and flow modelling. Due to increasing CO_2 dissolution with increasing pore flow velocity the area covered by gaseous CO_2 shrinks and the area affected by dissolved CO_2 expands at higher pore flow velocities. In total this effect leads to an enlargement of affected areas (Fig. 6.7) and thereby facilitates leakage detection.

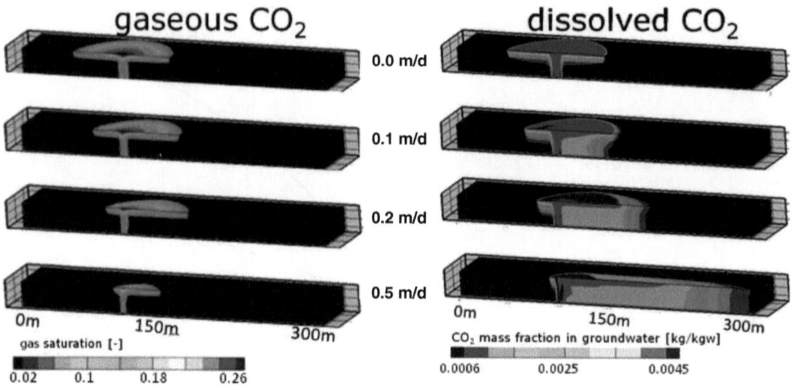

Fig. 6.6 Extensions of gaseous and dissolved CO_2 plumes under the influence of increasing pore velocities (0–0.5 m/day) in a homogeneous confined aquifer of 4 m thickness

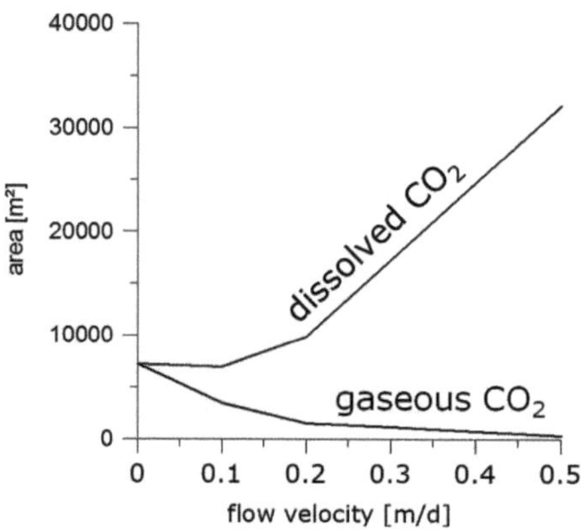

Fig. 6.7 Area affected by dissolved or gaseous CO_2. Calculated from results of Fig. 6.6

6.2.5 *Geophysical Surveys*

Geoelectrical investigations can measure the electric resistivity of the underground whereby boundaries between high resistivity sediments like sands and low resistivity units like clays or marls can be delineated. The electric resistivity is also strongly influenced by the electrical conductivity of the groundwater, so that changes in groundwater salinity can also be detected. Electric resistivity tomography (ERT) and a helicopter borne electromagnetic survey (SkyTEM) were conducted for two reasons. Firstly, much more information can be gained from such linear or areal investigations about the geology and hydrogeology than from punctiform depth profiles obtained from drillings. More detailed geological models can be developed on such data, which can help to optimize monitoring. And secondly, comparisons of repeat measurements could indicate changes in resistivity which can be either

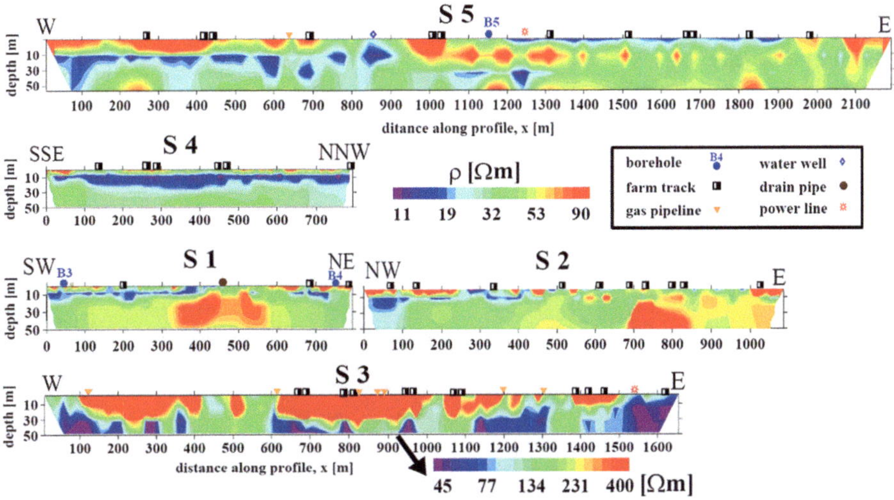

Fig. 6.8 Electric resistivity profiles measured from the surface

resistivity increases due to displacement of well conducting groundwater by a high resistivity CO_2 gas phase, or resistivity decreases caused by CO_2 dissolution which can be amplified by CO_2 driven mineral dissolution. Intrusion of highly conducting saline water would also be detectable. Both technologies were applied to test the capabilities of these methods and to obtain baseline measurements for the assessment of the intactness of the protected resource groundwater in case of CO_2 injections to the reservoir.

6.2.5.1 Geoelectrical Measurements (ERT)

In advance to the field measurements, a synthetic modelling study was conducted to compare different electrode configurations and to develop a setup optimized for the detection of thin (3–5 m) clay layers at depth down to 20 m. A multigradient configuration, which turned out from the modelling results to be most promising, was applied in the field. Five surface profiles with lengths of 800–2,240 m were measured with 160–448 electrodes of respectively 5 m distance (Fig. 6.8). Additionally borehole-surface configurations were measured at B3 and B4.

Since resistivities below 30 Ωm are typical for clayey sediments and resistivities above 50 Ωm are representative for sandy layers, the surface profiles S1, S2, S4 and S5 can be interpreted as heterogeneously interbedded clay and sand layers (Fig. 6.8). The exceptionally high resistivity in S3 is presumably caused by a series of gas pipelines in the underground which make the profile non-evaluable. A drain pipe crossing S1, may be also responsible for the high resistivity area below the pipe. Apart from these anomalies, distinct and extended geological structures can be identified, like for example a clayey layer in a depth between 5 and 30 m, which is prevailing in the western and northern part of the area. Since this layer seems to be relative continuously developed, it would cause spreading of CO_2 gas rising

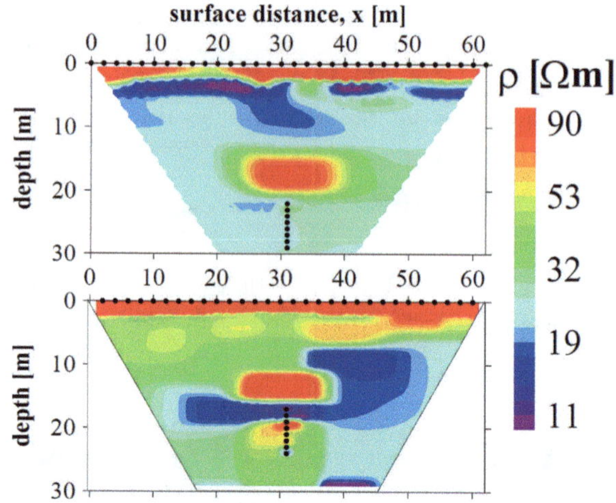

Fig. 6.9 Electric resistivity profiles from surface-borehole measurements

from the underground. With this knowledge about the location of such reduced hydraulic permeability layers, it is possible to optimize the position and depth of monitoring measures. Even more detailed information is provided by the borehole-surface measurements (Fig. 6.9).

6.2.5.2 Airborne Electromagnetics (SkyTEM)

To obtain an overview of the geology in three dimensions, an airborne survey covering the entire area was flown. The total number of line kilometres is 154 with a total number of 5,700 transient soundings.

The airborne survey was flown with the SkyTEM system (Fig. 6.10) which is a high powered absolutely calibrated transient electromagnetic system (TEM). TEM is well suited for locating conducting layers like clay and layers with salt water. The system transmits approximately 100 A in a four turn coil with an area of 314 m^2. The movement of the system in the airspace is monitored by GPS's, altimeters and inclinometers.

The measured data are processed in the Aarhus Workbench processing and inversion package (Auken et al. 2009; Sørensen and Auken 2004). This package corrects for the movement of the transmitter and receiver frame in the airspace. Infrastructure on the ground disturbs the data and therefore these data need to be culled. This is done partly automatically by special designed filters and partly by inspection of the raw data curves. In the final maps all artefacts from infrastructure are removed and resistivity variations reflect the geology only.

Average resistivity maps are shown for elevations −70 to −60 m and −170 to −160 m (Fig. 6.11). The system maps the geological formations. Low resistivities indicate clays or saline groundwater, intermediate resistivities are normally moraines while high resistivities are sand and gravel. The resistivity of sandy formations is highly dependent on the ion content of the ground water.

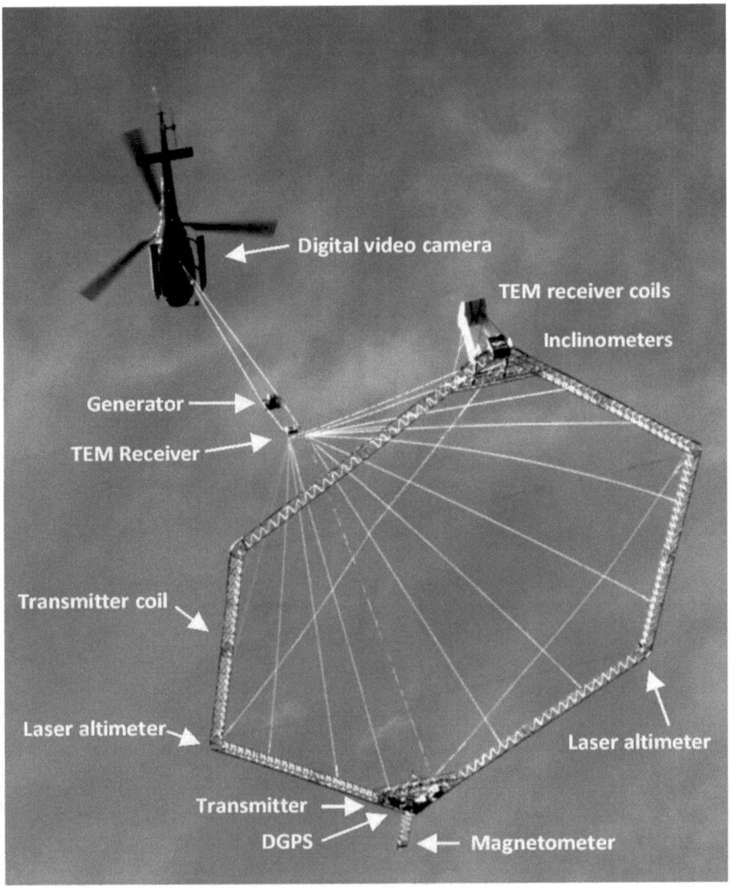

Fig. 6.10 The figure shows the airborne SkyTEM system in operation. The transmitter coil is attached to the lattice frame which also carries the altimeters, inclinometers etc

6.3 Vadose Zone Monitoring

Figure 6.12 clearly indicates several horizontal green bands in the uppermost 50 m, which most probably represent the boulder clay layers. Leaking CO_2 should accumulate under such structures and get visible there by repeating measurements. It is recommended to pay particular attention on such areas within the frame of groundwater sampling.

All environmental monitoring techniques point towards detection of possibly leaking CO_2 during or after an injection in a subsurface storage formation. The vadose zone is the final domain in the subsurface where migrating CO_2 can be identified before gas emission to the atmosphere occurs and subsequently can affect the environment and population. The aim of the vadose zone monitoring was to measure the concentration of CO_2 in the soil gas before (baseline), during and after

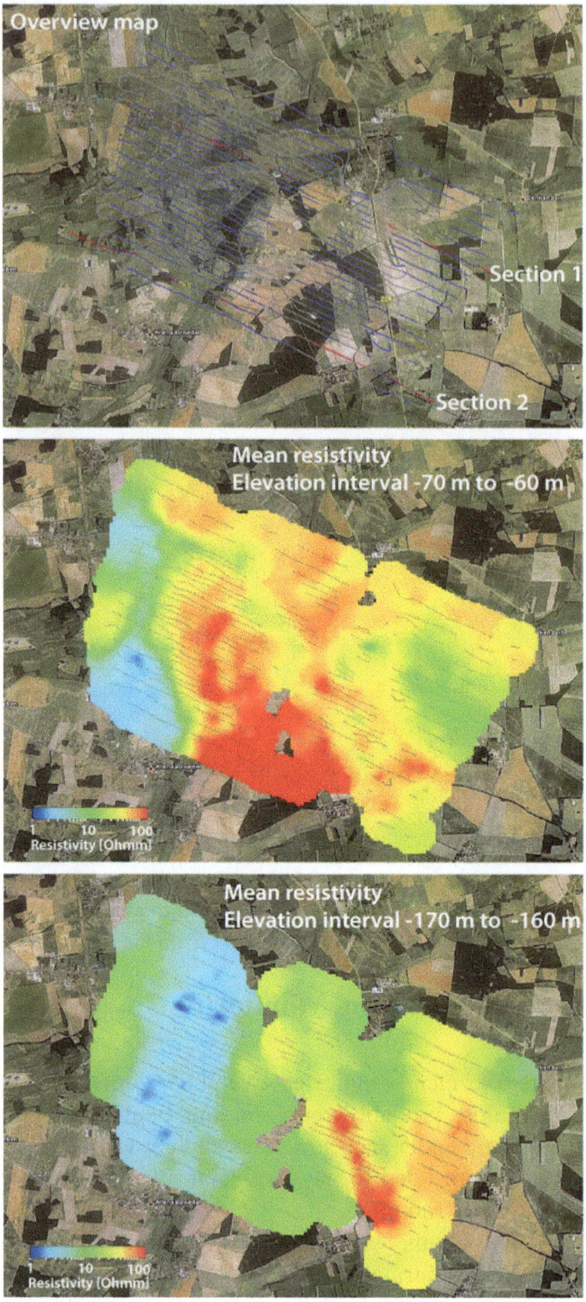

Fig. 6.11 The *top* map shows the flight lines. The two other figures show the average resistivity in the elevation interval −70 to −60 m and −170 to −160 m

an injection phase of CO_2 in the reservoir Altensalzwedel in the Altmark within the framework of an Enhanced Gas Recovery (EGR) project. Since the natural production of CO_2 by microbes in the upper soil is always present, the understanding of

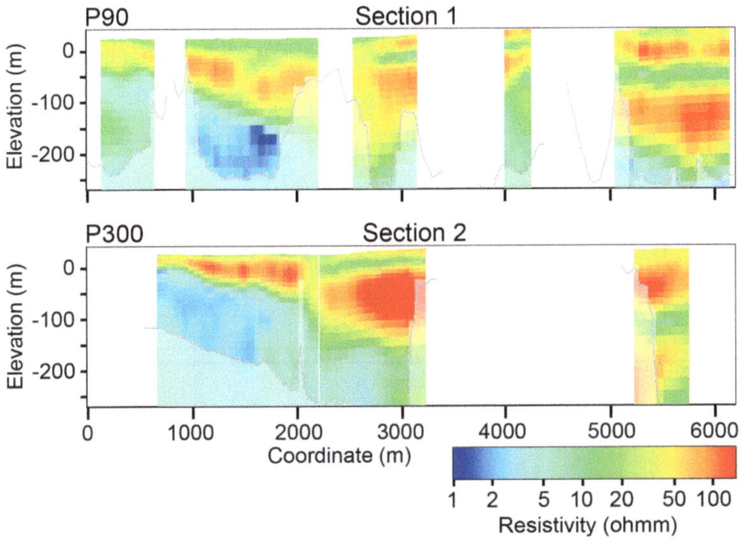

Fig. 6.12 The figure shows two cross sections striking east-vest. The whitening of the colours with depth indicates the depth of investigation. The interval between the two white lines should only be considered with care. The section should not be considered below the lowest white line

CO_2 soil gas concentration and its temporal variations is of vital importance to detect unnatural changes in the data.

The technical aim was to build up and run a network of continuously working monitoring stations and to optimize the instrumentation including the automatic data transfer.

In order to investigate the correlation of CO_2 variations with soil or meteorological parameters additional site specific measurements of various environmental parameters and the installation of a weather station completed the measuring programme. Due to the postponed CO_2 injection, no more than extended baselines could be measured and interpreted. But even in case of CO_2 injection a leakage into the vodose zone was not expected. Based on the results of 2 years monitoring an estimation of leakage sensitivity for all sites can be given.

6.3.1 Technical Setup

Main parts of the technical setup are displayed in Fig. 6.13. For site preparation gauge wells of 2 inches. in diameter were drilled and completed with a filter drain of half a meter length at the bottom and a tubing reaching up to the surface. Depending on local water levels at drilling time (autumn 2008) the measuring depths range from 1.15 to 3 m. The measuring principle is based on a closed gas circuit containing infrared gas sensors for CO_2 and CH_4, a sampling port, a pump, two

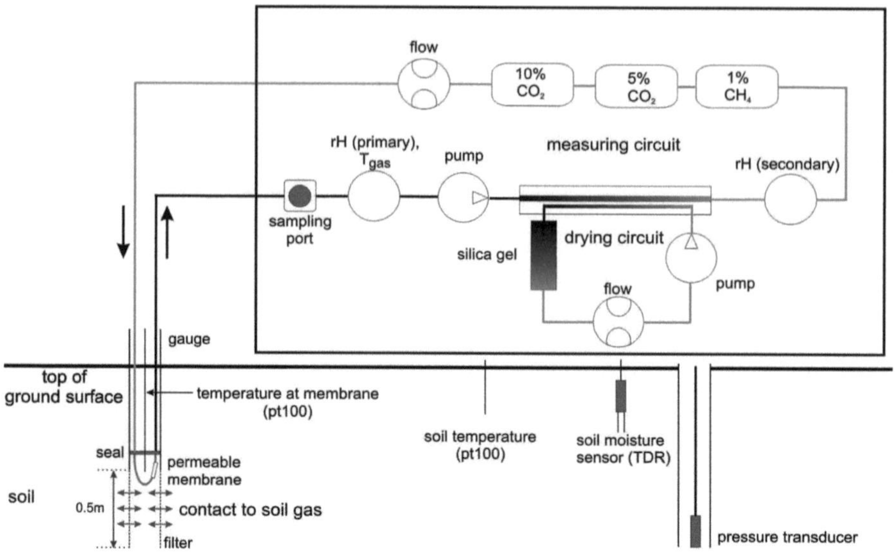

Fig. 6.13 Scheme of the technical design of the monitoring stations including additional sensors for soil and environmental parameters

gas humidity sensors and a flow sensor. The connection to the soil gas is realized by a gas permeable membrane leak-tight placed at the top of the filter drain. In order to protect the sensitive optic of the sensors, the gas has to be dried. For this purpose we use a drying tube connected to a second circuit including silica gel as a drying agent, a pump and a flow sensor. As additional parameters soil moisture, soil temperature, temperature at depth of the membrane, air temperature and water level in a second well using a pressure transducer were recorded at each site.

The measured data are locally stored on two programmable data loggers in 3 h files, which are automatically transmitted via the Global System for Mobile Communications (GSM) to the ftp server of the Federal Institute for Geosciences and Natural Resources (BGR) in Hannover. All data are stored in a special data base (SQL Server). For an easy to use data request and graphical display a JAVA application was developed.

To get the regional meteorological information a weather station containing sensors for air temperature and air humidity, wind speed and direction, the atmospheric air pressure, the precipitation and the global radiation was installed at one site.

6.3.2 Area of Investigation

Figure 6.14 shows a detail of the topographic map including the outline of the reservoir as a black line. In total 12 monitoring stations were placed next to

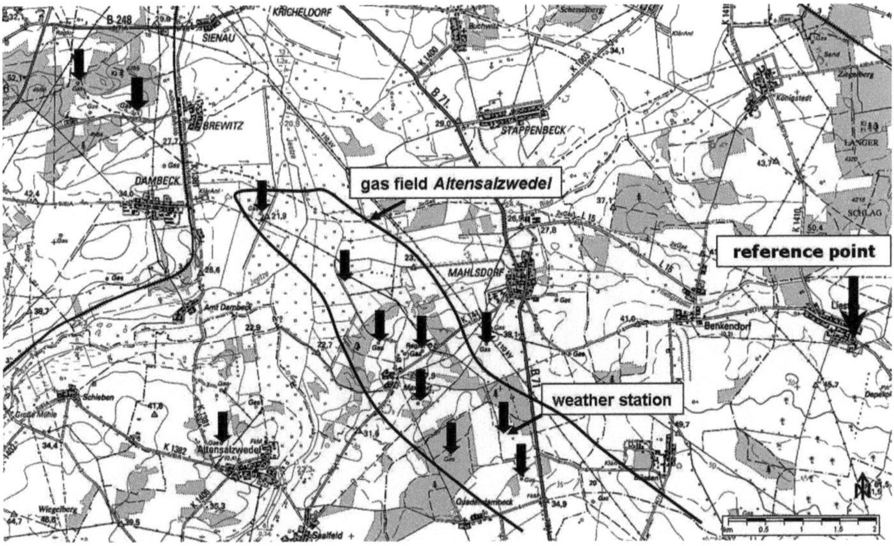

Fig. 6.14 Detail of the topographic map of the investigation area including a surface projection of the reservoir and the position of the stations

different well sites (about 20 m apart of the well head) including the two planned injection wells. The subfield is structurally and hydraulically isolated from the main field Peckensen and as a natural gas reservoir the structure itself can be described as hermetically sealed. Therefore the partly 40 years old production wells stand in the focus of investigation as points for a potential leakage. The sites, denoted by arrows, are located above or in the vicinity of the Altensalzwedel gas field. One additional station was installed about 5 km east of the reservoir in the village "Liesten" as a reference point. The weather station is placed in the southeast of the investigation area at one well site lying in the centre of the monitoring area with the best local conditions for meteorological observations.

6.3.3 Measurements and Interpretation

The individual baselines at the different sites show a wide interval of observed maximum CO_2 concentrations ranging from 0.2 % to more than 19 %. Figure 6.15 shows a comparison of seasonal (winter/summer) maximum CO_2 concentrations at all 13 sites. The lower part of the figure displays the corresponding filter depth where the actual soil gas concentration is measured. As mentioned above the varying installation depth of the filter results from the different local water levels while drilling (denoted by the horizontal lines in Fig. 6.15) the gauge wells. Hence, it was not possible to realize an identical configuration at all sites.

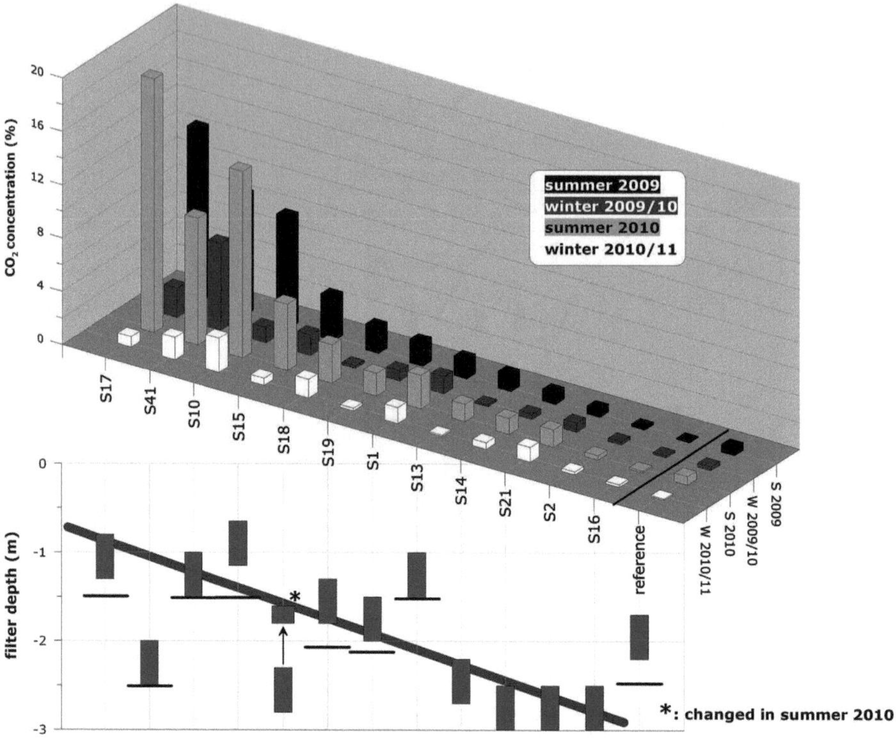

Fig. 6.15 Seasonal maximum concentrations of CO_2 in the soil gas at all sites, corresponding filter depth and the water level at drilling time

The monitoring of water levels shows a very strong temporal variation at some sites. The tail water flooded the filter part for several weeks or months at some of the stations. Consequently during these time periods the concerned stations do not record actual soil gas concentrations but equilibrium concentrations between dissolved gas in the tail water and the gas stream. These "data gaps" were omitted for further interpretation of soil gas variations.

From the data it is obvious, that a negative correlation between CO_2 concentration and measuring depth exists (highest CO_2 concentration at lower depths). This is certainly due to the enhanced microbiological CO_2 production at shallower depths (microbiological active zone). Deviations of this general trend, e.g. the site S41, can be reasonably explained with particular soil characteristics. Investigations of well core samples revealed an elevated content of organic carbon close to the measuring depth and high potential CO_2 production rate of microorganisms.

Principal seasonal variations show higher values in the summer and lower in the winter. The time series of the CO_2 concentration is primarily influenced by at least two parameters: the soil temperature and the soil moisture. In autumn and winter the CO_2 concentration follows more or less the temperature curve while the soil

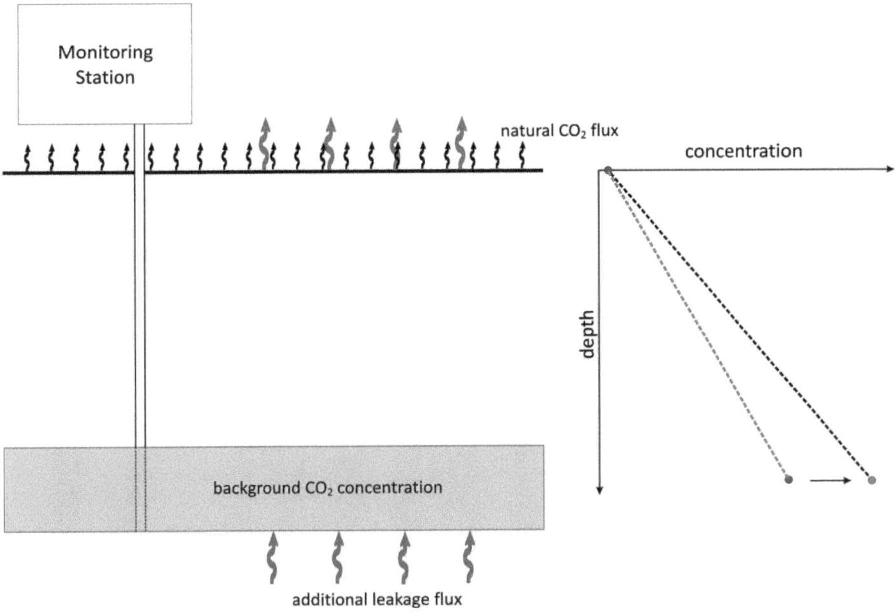

Fig. 6.16 Scheme of leakage scenario regarding an additional upward migrating (diffusive) laminar flux which causes a concentration increase in the observed soil layer

moisture is always high during that time. In spring the CO_2 concentration increases and reaches a maximum in June.

6.3.4 Potential Leakage Detection

To estimate the required contribution of CO_2 to be able to detect a potential leakage in the vadose zone a simplified model was calculated. Figure 6.16 displays the definition of the modelled situation. The monitoring station records the background CO_2 concentration in a distinct soil layer. The CO_2 distribution in the soil leads to a natural background flux through the surface into the atmosphere. Assuming an additional leakage diffusive flux from depth the gas will migrate through the soil layer under investigation before getting through to the surface.

Chamber measurements in August 2010 revealed an average CO_2 flux at the surface of about 200 kg/(day*ha) in the Altmark area. As an exemplary leakage scenario an additional CO_2 gas emission at the surface of 50 kg/(day*ha) was assumed caused by the migration of gas from a deeper interval. As a result the CO_2 concentration gradient will increase by approximately 0.8 % per metre depth. This "dispersive leakage scenario" will lead to a very slow increase of the soil gas concentration rather than the expulsion of CO_2. This diminutive increase will not be detected on short measuring intervals (or discrete soil gas surveys) as the natural variation of CO_2 due to diurnal/seasonal variations and meteorological conditions

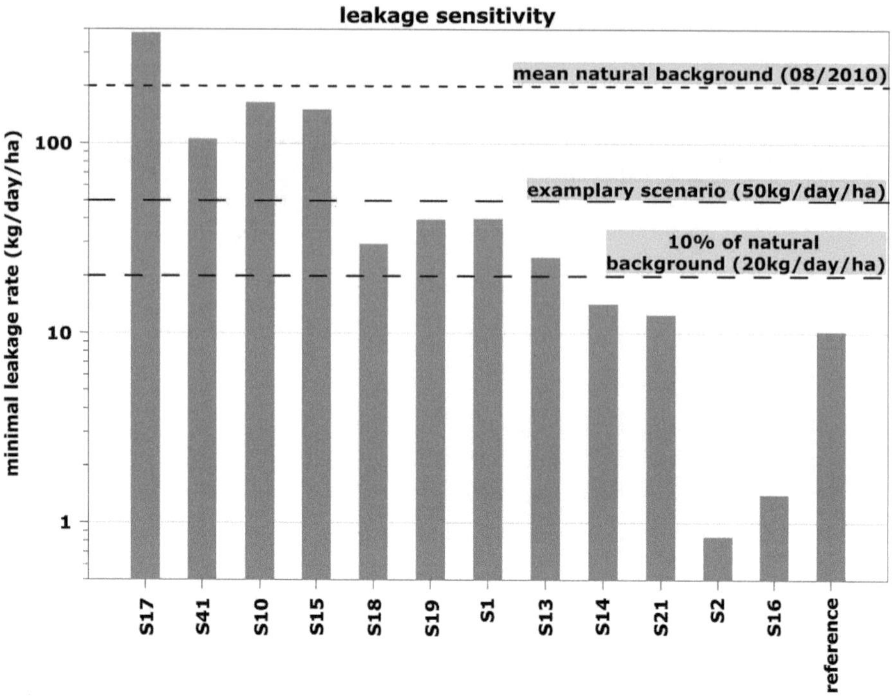

Fig. 6.17 Minimal leakage rate to be outside the 2σ interval as a degree of leakage sensitivity at all sites

will superimpose the additional flux. Hence the long-term average values were compared with the described leakage scenario.

The mean value of the CO_2 concentration and the standard deviation is different for all sampling sites. Some of the monitoring sites reveal a large variation in gas concentrations. However, the sensitivity of each site for leakage detection needs to be estimated by calculating the expected CO_2 increase in soil gas concentration by combining the gradient increase (0.8 %/m) with the measuring depth. The computed CO_2 concentration rise is then added to the mean value of each station. Doing that, the leakage sensitivity for each site can be calculated (Fig. 6.17).

For comparison the value of the mean natural background flux, the exemplary scenario and 10 % of natural background flux are added. Based on that the sites can be classified into three groups:

- Very low sensitivity at station S17 (here the minimal leakage rate has to be nearly twice as high as the natural flux), S41, S10 and S15. For the purpose of monitoring of a CO_2 reservoir they are hardly suitable.
- Medium sensitivity at stations S18, S19, S1 and S13. With the additional information of environmental and meteorological parameters they provide a reasonable compromise.

- High sensitivity is shown by the stations S14, S21, S2, S16 and the reference station, with nearly ideal conditions at the sites S2 and S16.

It can be summarized, that soil gas monitoring can be a suitable and sensitive method to detect non-natural CO_2 concentrations in the vadose zone. The preconditions are the possibility to measure the CO_2 concentration significantly below microbiological active zones but above varying ground or tail water levels. Both constraints have to be tested by preliminary site investigations. Because of the strong local and seasonal variations individual baselines for at least 2 years are necessary for interpretation of the CO_2 concentration measurements.

6.4 Atmospheric CO_2 Release and Airborne Dispersal

The underground storage of carbon dioxide and the usage of CO_2 for increasing the gas yield from natural reservoirs (enhanced gas recovery – EGR) are associated with potential risks of CO_2 leakage and release, especially due to the handling of CO_2. The release of CO_2 from the storage complex could occur into the groundwater or even into the atmosphere.

For the permitting viability of CO_2 storage and EGR projects, as well as for public acceptance, the depiction of safe contact with these gases plays an important role. A related focus is the description of uncontrolled, accidental releases, the subsequent gas dispersal processes in the atmosphere, and the resulting potential hazards for humans.

The current public debate gives evidence that such hazards to humans are taken seriously. With respect to examples such as the CO_2 release from Lake Nyos crater lake or the Lazio Region (IZ Klima 2010) and also the CO_2 incident in Mönchengladbach in August 2008 (Lampe and Schattka 2009), possible hazards will be discussed. This also shows that the debate is framed by just a few examples and scarcely through systematic scientific investigation. The particularities of the individual accidents in the CCS literature are as a rule not separately described, but rather commonly referred to as general CO_2 accidents.

The safe handling of CO_2 is accomplished through current technology and is standard practice in a multitude of facilities. Safety data sheets and handbooks are available (e.g. EIGA 2009; Linde 2003; Air Liquide 2010) in which the way of safe handling of CO_2 is described. The difference to CO_2 storage and EGR results from two important criteria:

- Significantly larger amounts of CO_2 are handled and
- CO_2 is spent underground and (apart from surface facilities) not handled in a technical installation (e.g. tanks), where it can be significantly easier controlled.

Based on these aspects, extensive requirements result for systematic risk assessment and risk management. Within the framework of risk management, impact prognoses of CO_2 releases into the atmosphere are of primary importance. The impact is composed of complex release and dispersal processes. For the description of these processes, the basic resources are already available, but with respect to CO_2 storage and EGR, these processes have not been given the same level of attention as, for example the storage reservoir modelling.

6.4.1 CO_2 Release Scenarios

CO_2 can reach near-surface compartments via various pathways. Depending on the scenario, various compartments are thereby affected, which on the one hand can cause an adverse effect, but on the other hand present an opportunity for detection.

In the CO_2 storage literature, CO_2 release is often reduced to release only in the immediate vicinity of the storage area. Within the framework of comprehensive EGR and CO_2 storage consideration, above-ground facilities are to be regarded as well. These facilities belong indeed, to the storage complex according to definitions from Guideline 2009/31/EG of the European Parliament from 23.04.2009 on the geologic storage on carbon dioxide (EU CCS Directive 2009), but have played to this point only a secondary role in the risk assessment literature. The above-ground facilities can, however, play a decisive part, especially in terms of short-term release to the atmosphere. The relevant scenarios for a CO_2 release to the atmosphere can be represented as follows:

- Release from the storage complex:
    - Release via active well bores,
    - Release via abandoned wells,
    - Release due to advective transport from the reservoir via existing or otherwise caused pathways (e.g. faults);
- Release from above-ground facilities:
    - Leaks in storage and transport facilities,
    - Release during filling of above-ground facilities (e.g. filling of storage tanks for intermediate storage of CO_2),
    - Release during injection,
    - Release from above-ground safety installations, for example blow-out systems.

6.4.2 Calculation of Gas Dispersal in the Atmosphere

CO_2 is a non-toxic, colourless and odourless gas. Organoleptic measures for detection are rendered therefore ineffective. Solely through condensation of air moisture, resulting in a white water vapour fog, can a CO_2 release be visible. Its presence and colouring are, however, extremely dependent on air humidity and release conditions and provide no conclusions to the mass of the CO_2. CO_2 in deadly concentrations can be absolutely invisible.

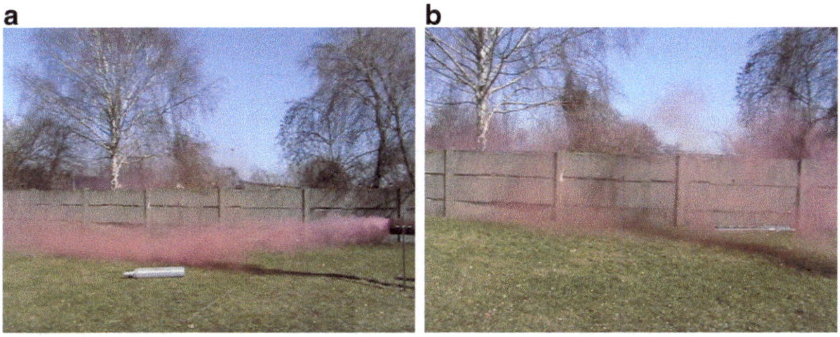

Fig. 6.18 CO_2 immediately after release from a leak (*left*) and CO_2 swirling in the wind (*right*)

A method for visualization is currently under development (Großmann et al. 2010). Released CO_2 is made visible by colouring, allowing the visual detection of a potential hazard (Fig. 6.18). With such a method, a very important disadvantage in CO_2 risk management is overcome – the hazard can be made optically perceptible. Such a method is simultaneously a useful tool for conducting further research on CO_2 dispersal.

Both types of release (Fig. 6.18) and the underlying mechanisms (Fig. 6.19) are important for the dispersal calculation. Important release mechanisms to be considered are:

- Free-jet dispersal (with and without influence of crosswind),
- Direct, impulse-free gas dispersal,
- Gas dispersal with sublimation source terms,
- Gas dispersal following heavy gas discharge,
- Impulse-affected release of CO_2 from the liquid phase.

The calculation of gas release and dispersal follows the following fundamental equations:

- Real gas model of Redlich/Kwong,
- Gas dynamics relating to conservation of mass,
- Impulse and energy theorems of viscous fluids.

6.4.2.1 Applied Methods

Heavy and light gas dispersal: The dispersal processes for heavy gas was studied according to VDI Guideline 3783 "Dispersal of Accidental Releases – Safety Analysis, Sheet 2" (VDI-RL 3783, incl. authorized software package P&K 3783). The calculation of heavy gas discharge was conducted with the "shallow water" flow model, which is driven by the gravitational field of the earth and the density difference between the heavy gas stream and ambient air. The specific topography and the influence of structures were taken into account. The calculation method was

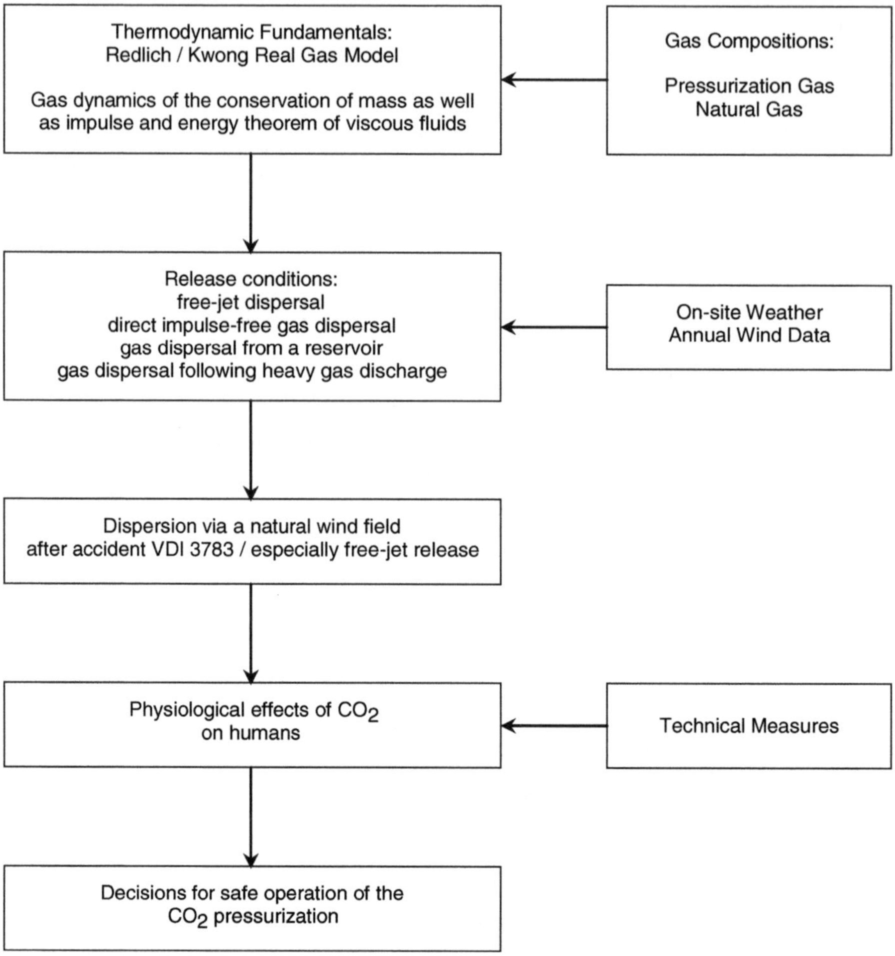

Fig. 6.19 Flow chart of the calculations for gas dispersal and assessment

further developed and applied (Großmann and Naue 2009). The calculations for light gas dispersal were performed according to VDI Guideline 3783 as well.

Free-jet calculation methods with influence of crosswind: To date there are no standardized calculation methods available for free-jet dispersal under influence of crosswinds. In the calculations presented here, own methods were used, which are based on experimental investigations (Naue 2001) and proven software (Großmann and Naue 2009).

To date, there are no existing, completed investigations of CO_2 dispersal. Within the framework of the CLEAN project, several release and dispersal scenarios were systematically investigated (Großmann and Naue 2009). This demonstrated that a comprehensive R&D need for verification and derivation of easy-to-manage calculation programmes for defined release and dispersal scenarios still exists.

6.4.3 *Calculation of Example Release and Dispersal Scenarios*

6.4.3.1 Aboveground Facility Releases

Blow Out via a Safety Valve

The pressure release valve of a storage tank or other facility components secured by safety valves (blow out system) takes place by opening a safety valve during a failure of normal operation. The safety valves respond promptly before reaching the highest permitted pressure in the storage tank. Safety valves can be installed in a horizontal or vertical manner. From a safety point of view, advantageous is normally a vertical installation.

The geometry of the blow out line from the storage tank through the narrowest diameter of the safety valve to the outlet cross section and the thermodynamic state variables in the storage tank together determine the discharge parameters of the free-jet in the vicinity.

The narrowest diameter of the safety valve defines the mass flow rate during pressure release. During the gas-dynamic investigation of the flow through the pressure release line, the real gas behaviour (compressibility factor) is to be taken into account. Furthermore, during an adiabatic pressure release in the storage tank, it is required that the specific enthalpy, which is displayed through the resting pressure and resting temperature on the gas side in the storage tank, does not change during the release of pressure. The material transfer between liquid phase and gas phase during the pressure release was neglected. The initial gas volume in the storage tank therefore remains constant during the course of the pressure release.

As an example, a calculation for the pressure release of a CO_2 buffer tank was carried out, as it could be a component of an aboveground facility. For this example, an initial mass flow rate of 42.17 kg/s CO_2 resulted (Großmann and Naue 2009). In Fig. 6.20 calculation results for the blow out of CO_2 via a vertical blow out valve for the abovementioned example are shown. CO_2 concentrations of 0.5 and 4 vol.% were used as assessment criteria. From the calculated and displayed examples (Fig. 6.20), the following statements were derived:

- The detailed investigations of free-jet dispersal via both variants of the blow out valve configuration display the following maximum extent of the hazard zones for an assessment value of 4 vol.% CO_2:
  - Calm winds: at a height of 10 m to a distance of 27 m,
  - For crosswinds of up to 25 m/s, the distance is reduced to 3 m,
  - Near to the surface, hazards can be present up to a maximum distance of 25 m.
- For vertical blow out systems, sufficient dilution to below 0.5 vol.% near the surface is ensured in every case. A hazard for people located at the surface can therefore be ruled out with vertical blowout systems.

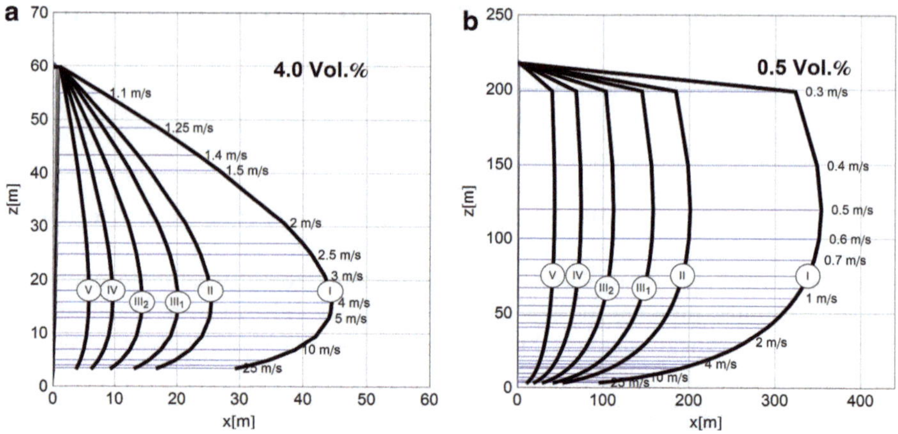

Fig. 6.20 Free jet axes (*thin*) and hazard zones (*thick*) by different atmospheric turbulence classes (I, II, III1, III2, IV, and V) and wind forces (0 to 25 m/s) for vertical blow out. Assessment criteria 4 % CO_2 (*left*). Assessment criteria 0.5 % CO_2 (*right*)

- Another significant advantage of the vertical blow out system is the impossibility of impulse destruction of the free jet from encountering of obstacles. Therefore, a related hazard to people can also be ruled out.
- The most important hazard for horizontal blow out systems is the impulse destruction of the free jet upon encountering an obstacle (e.g. people, trees, or buildings). Significant direct impulse impacts on people and significant sound levels and consequently hazards to people in the vicinity of the blow out system are possible.
- Proper configuration of the blow out system can safely prevent hazards outside of the operational premises.

6.4.3.2 Release from Storage Complex

Releases from the storage complex can occur suddenly and eruptively, e.g. resulting from well bore failure and reflect in this case a point source. However, leakage from a larger area could occur as well, if CO_2 passes the direct cap rock of the reservoir and afterwards the multi barrier system consisting of further reservoirs and cap rocks, finally reaching the surface through the topmost groundwater aquifer and the soil gas area. Whether this kind of leakage results in a rapid release of larger amounts of gas via preferential flow paths or whether the CO_2 first dissolves in the groundwater and is then released, depends on site specific conditions.

A long list of CO_2 releases from CO_2 injection well bores is documented in literature. A good summary can be found in Skinner (2003). In this article, on average two CO_2 blow outs per year for the last 27 years are described. These deal with blow outs during CO_2 injection for the purpose of increasing production rates in oil and natural gas fields (enhanced oil/gas recovery EOR/EGR). It is also explained in Kretzschmar et al. (2008) that the number of eruptive occurrences from wells declined in recent years, however still happen.

Reasons for these CO_2 releases are primarily handling errors and corrosion. Acting against both causes, sufficient preventive measures have been enacted. The corrosion problem can, however, be complicated with regard to old and abandoned bore holes. Alongside primary measures for release prevention, additional comprehensive actions can be taken for early detection of leaks and consequently for reduction of eruptive release of large amounts of gas.

Until now, very few simulations of eruptive releases of CO_2 from bore holes into the atmosphere have been made available. In order to be able to estimate the effects of an eruptive release from a well considering the release of a two-phase mixture and the possible formation of a snow/ice pool, an exemplary investigation was conducted (Großmann and Naue 2010). As a worst case, it was assumed that the release resulted from demolition of the Christmas tree and simultaneous failure of the safety valve. The following boundary conditions were assumed herein:

- Well head temperature of the fluid: −30 °C to 0 °C,
- Well head pressure of the fluid: 2.0–3.0 MPa,
- Temperature of the fluids at depth of 3,250 m: 60–100 °C, approximately linear temperature increase with depth,
- Diameter of the damage to the Christmas tree: 7″-line, resulting in an inner diameter of 157 mm.

Within the scenario the release took place in two stages:

1. Release of a two-phase fluid,
2. Subsequent release of a liquid phase.

The release of the two-phase mixture is primarily determined by the well head temperature. At 273 K, a mass of CO_2 (two-phase) of 8.4 t was calculated, which was released in ca. 12 s. This corresponds to a mass flow of ca. 640 kg/s for the first 7 s (corresponding to approximately 2,300 t/h). In the subsequent release time, the CO_2 mass flow was calculated to be at 1.3 kg/s (= 4.7 t/h or 112 t/day) with regard to the assumed boundary conditions.

For the assumed demolition of the Christmas tree in the present example, the chronology of the release occurred as follows:

- Upon demolition of the Christmas tree and failure of the safety valve, which is located at a depth of approximately 40 m, a low pressure wave flows downward and enlarges an existing – or produces a new, two-phase fluid to a defined depth, where the boiling pressure is equal to the current local pressure. In the horizontal pipe, a seal is assumed.
- If the entire horizontal pipe is also enveloped by the pressure wave, this would result in significantly larger release volumes. The damage to the horizontal pipe has already been examined during a gas or liquid release, but not for a two-phase fluid.

- The two-phase fluid flows out as dispersion with a critical velocity in open pipe cross-sections and as a subsequent post-expansion with super-critical velocity as a vertical free jet. The critical velocity of the dispersion results from the examinations of maximum mass flow density.
- The spatially limited two-phase fluid flows out, dependent upon the initial parameter constellation, in 10–14 s and leaves behind a liquid in thermal equilibrium at boiling temperature at the interface of the two-phase fluid and the liquid.
- The liquid phase then flows up into the well head of the injection well with a velocity of approximately 0.082 m/s and causes a release mass flow of 1.3 kg/s, once the cooling process of the back-flowing liquid mass in the well bore pipe takes place at the same rate as its heating during downstream flow under normal operational conditions.
- The low pressure wave is reflected from the reservoir at depth as an overpressure wave with the same amplitude and arrives at the damaged location after approximately 20 s. In the event that the area where the damage occurred is not yet frozen, the overpressure wave coming up from below is reflected downward as an underpressure wave and triggers the same process again. If the site of the demolition is frozen in the sense of self-repair, the overpressure wave coming up from below will be reflected downward as an overpressure wave with doubled amplitude. In this context, ice plugs can be completely pushed out of the pipe. Repeated pressure wave reflections are therefore not to be ruled out.
- Possible release of CO_2 in pulsation cycles resulting from the pressure wave reflection process.

Within the investigations it was assumed that, on the one hand the gas portion of the two-phase fluid disperses as a free jet independent of the fluid/solid phase and on the other hand the fluid/solid phase moves at the same starting velocity in the gravitational field of the earth.

6.4.3.3 Release of the Fluids via Blast Plumes

Free jets in crosswind have curved blast plumes. If the average velocity in the free jet reached the velocity of the horizontal crosswind, horizontal plume dispersal follows. The regularity of the dilution in the free jet (inversely proportional to the covered free jet pathway) is differentiated from that in the plume (inversely proportional to the square of the distance from the jet due to crosswind). The plume dilution is dependent on the turbulence of the wind field. The turbulence of the wind field is categorized through dispersal classes 1–6. The combination of free jet and dispersal plume is known as a jet plume.

The extent of the CO_2 with hazardous concentrations was assessed based on the value 4 vol.% CO_2. For this release case, there are no near-surface ($z < 10$ m) hazards as a consequence of the jet plume dispersal at a horizontal distance of greater than 36 m to be expected.

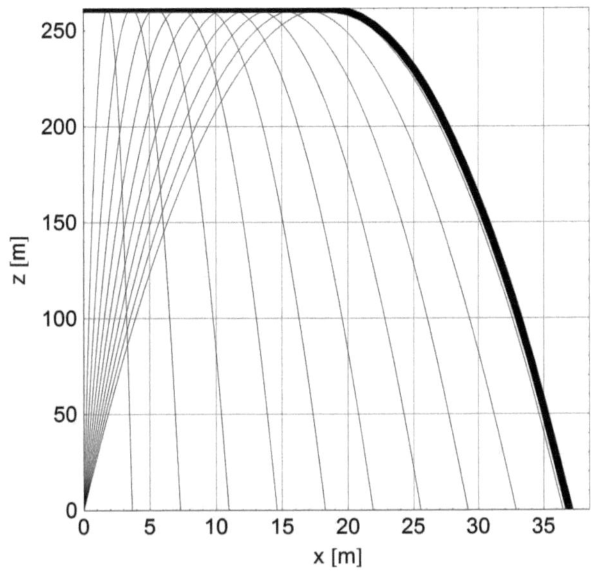

Fig. 6.21 Trajectory parabola with start angle against the horizontal in the interval $88° \leq \alpha \leq 90°$

6.4.3.4 Release of the Fluid as Liquid Eruption

The hypothesis that the liquid portion of the dispersion moves upwards or laterally upwards with a certain velocity and then falls back was tested. Therefore, the subsequent evaporation from the pool was calculated. This was done for the first pulsation cycle. Either the pool remains with this first cycle or the pool increases its height due to the subsequent similar pulsation cycles. This only has influence on the release time, not however on the quasi-stationary concentrations in the vicinity.

If the entire mass of the liquid during the eruption occurring after demolition of the Christmas tree and failure of the safety vent is regarded as liquid with an escape velocity of 71.6 m/s, then trajectories, for different launching angle result (Fig. 6.21).

Both, the precipitation in a pool and the subsequent evaporation and the complete evaporation of the liquid phase were examined. The following scenarios were examined applying VDI-Guideline 3783:

1. Stationary evaporation from a pool with a radius of 36 m with a mass flow density of 0.31 g/(s m^2) resulting from solar radiation and concurrent heat transfer of 1,200 W/m^2. The mass flow amounts to 12.62 kg/s. The complete evaporation of the pool with a mass of 4,000 kg takes place in the release time of 317 s. Due to the possible cyclic repetition of the eruption until freezing at the location of the leak, the process is considered as stationary contamination of the vicinity.
2. Dispersal of a gas mass of 4,400 kg in an initial volume with a surface area of 4,071 m^2 and an assumed layer thickness of 50 m (average mass density of 21.6 g/m^3 or 1.1 vol.% CO_2).

Upon a release from the pool, a hazardous area of approximately 175 m therefore results during unfavourable dispersal conditions. For this, a level surface for the dispersal was provided. For sloping terrain, even larger hazard areas can result during unfavourable circumstances in the direction of the gradient. The advantage is, however, that the gradient situation is known and monitoring and intervention measures can be foreseen in the presumed dispersal direction.

Overall, the calculations show that no blanket statements as to the hazardous areas resulting from eruptions from a bore hole are possible. The possible hazardous situations must be determined in a much more event- and location-specific manner. Based on this approach, however, site-specific prognoses and thus the introduction of specific monitoring and intervention measures are possible.

References

Air Liquide (2010) Safety data sheets as per guideline 1907/2006/EG (REACH), Current as of 13.10.2010

Auken E, Christiansen AV, Westergaard JH, Kirkegaard C, Fopged N, Viezzoli A (2009) An integrated processing scheme for high-resolution airborne electromagnetic surveys, the SkyTEM system. Explor Geophys 40:184–192

EU (2009) CCS Directive 2009/31/EG of the European Parliament and of the Council of 23 April 2009 on the geologic storage of carbon dioxide

Großmann J, Naue G (2009) Expert opinion on malfunctions during CO_2-RECOVERY in Altmark mining area. Internal, unpublished report, GICON

Großmann J, Naue G (2010) CO_2 release resulting from borehole damage. Internal, unpublished report, GICON

Großmann J, Schreck A, Cotta F (2010) Tests on CO_2 dispersal. Internal, unpublished report, GICON

Großmann J, Schilling F, Dahmke A (2011) Zusammenfassende Darstellung zur Sicherheit und Umweltverträglichkeit der CO_2-Speicherung, in Sicherheit und Umweltverträglichkeit der CO_2-Speicherung. UBA – Umweltbundesamt, p 76

EIGA (European Industrial Gases Association AISLB) (2009) CO_2 Tanker Driver Manual, http://www.eiga.org/, IGC Doc 56/09/E

IZ Klima (2010) Speicherung von CO_2: Technologie für den Klimaschutz. Open publication, http://www.iz-klima.de/

Köber R, Dahmke A (2011) Ableitung eines standortspezifischen Monitoringkonzepts für die CO_2-Speicherung im Rahmen einer EGR- oder CCS-Maßnahme. Institut für Geowissenschaften, Christian-Albrechts-Universität zu Kiel, Kiel, p 96

Köber R,Wiegers C, Schäfer D, Dahmke A, Großmann J, Tischer L, Beyer M (2011) Ableitung eines standortspezifischen Langzeit-Monitoringkonzepts für EGR unter Einbeziehung der rechtlichen und sicherheitstechnischen Vorgaben sowie hydrogeochemisches Monitoring im flachen Grundwasserleiter. Endbericht der Teilprojekte Mo1 und Mo3 im BMBF-Forschungsvorhaben CLEAN. Institut für Geowissenschaften, Christian-Albrechts-Universität zu Kiel, Kiel, 414 p.

Kretzschmar H-J, Lubenau U, Müller-Syring G (2008) Expert system CO_2 borehole integrity for the "COSMOS-Technology of CO_2 storage, monitoring and safety" project as part of the BMBF/DFG Program GEOTECHNOLOGIEN, Freiberg

Lampe J, Schattka, D (2009) Brandeinsatz mit Kohlenstoffdioxidfreisetzung in einem Lagerbetrieb. Brandschutz - Deutsche Feuerwehrzeitung (9), p 747–753.

Linde AG (2003) Safety guidelines for contact with CO_2. Document 12/03.03/8459/0

Naue G (2001) Free jet streams in crosswinds: contributions to fluid mechanics. Technische Universität Dresden, pp 245–252, ISBN 3-86005-276-4

NETL (2009) Monitoring, verification, and accounting of CO_2 stored in deep geologic formations. NETL – National Energy Technology Laboratory, Albany, p 132

Parkhurst DL, Appelo CAJ (1999) User's guide to PHREEQC (version 2) – a computer program for speciation, batch-reaction, one-dimensional transport, and inverse geochemical calculations. Water-Resources Investigations report 99–4259, U.S. Geological Survey, Denver

Peter A, Hornbruch G, Dahmke A (2011) CO_2 leakage test in a shallow aquifer for investigating the geochemical impact of CO_2 on groundwater and for developing monitoring methods and concepts. Energy Procedia 4:4148–4153

Skinner L (2003) CO_2 blowouts: an emerging problem. World Oil Magazine 224(1):38–42

Sørensen KI, Auken E (2004) SkyTEM – a new high-resolution helicopter transient electromagnetic system. Explor Geophys 35:191–199

Spangler LH, Dobeck LM, Repasky KS, Nehrir AR, Humphries SD, Barr JL, Keith CJ, Shaw JA, Rouse JH, Cunningham AB, Benson SM, Oldenburg CM, Lewicki JL, Wells AW, Diehl JR, Strazisar BR, Fessenden JE, Rahn TA, Amonette JE, Barr JL, Pickles WL, Jacobson JD, Silver EA, Male EJ, Rauch HW, Gullickson KS, Trautz R, Kharaka Y, Birkholzer J, Wielopolski L (2009) A controlled field pilot for testing near surface CO_2 detection techniques and transport models. Energy Procedia 1:2143–2150

Spangler LH, Dobeck LM, Repasky KS, Nehrir AR, Humphries SD, Barr JL, Keith CJ, Shaw JA, Rouse JH, Cunningham AB, Benson SM, Oldenburg CM, Lewicki JL, Wells AW, Diehl JR, Strazisar BR, Fessenden JE, Rahn TA, Amonette JE, Barr JL, Pickles WL, Jacobson JD, Silver EA, Male EJ, Rauch HW, Gullickson KS, Trautz R, Kharaka Y, Birkholzer J, Wielopolski L (2010) A shallow subsurface controlled release facility in Bozeman, Montana, USA, for testing near surface CO_2 detection techniques and transport models. Environ Earth Sci 60(2):227–239

Xu T, Sonnenthal E, Spycher N, Pruess K (2006) TOUGHREACT-A simulation program for non-isothermal multiphase reactive geochemical transport in variably saturated geologic media: applications to geothermal injectivity and CO_2 geological sequestration. Comput Geosci 32 (2):145–165

Chapter 7
Theoretical Test Case of the Injection of 100,000 t of CO_2 into the Altmark Depleted Gas Field

Michael Kühn, Robert Meyer, Kilian Nchungong Awemo, Michaela Bock, Stefan Buske, Leonhard Ganzer, Rüdiger Giese, Lars Houpt, Juri Perestjuk, Erhard Ribbe, Gerhard Rosenthal, Steffen Schmitz, Cori Schreyer, Lutz Stecken, Arron T. Singhe, Matthias Tantow and Gernot Voigtländer

Abstract CLEAN was a scientific programme in support of a pilot Enhanced Gas Recovery (EGR) project and it was planned to inject nearly 100,000 t of carbon dioxide (CO_2) into the Altmark natural gas field. Due to delays in the permitting process, injection did not occur within the time frame of the project. Therefore a test case was studied of the theoretical injection of CO_2.

Modelling CO_2 injection in the Altensalzwedel segment of the Altmark depleted gas field shows that effective injection can be carried out at a tubing head pressure as low as 3.5 MPa and at tubing head temperature of 10 °C.

The history matching of the simulation model and the injection prediction were successfully conducted with a volume ratio of injected CO_2 to the gas in place of approximately 0.06 which did not show any enhancement in the gas

M. Kühn (✉) • M. Bock • R. Giese
Helmholtz Centre Potsdam, GFZ - German Research Centre for Geosciences, Telegrafenberg, 14473 Potsdam, Germany
e-mail: michael.kuehn@gfz-potsdam.de

R. Meyer • J. Perestjuk • E. Ribbe • G. Rosenthal • C. Schreyer •
L. Stecken • M. Tantow • G. Voigtländer
GDF SUEZ E&P Deutschland GmbH Waldstrasse 39, 49808 Lingen, Germany

K.N. Awemo • L. Ganzer • A.T. Singhe
Institute of Petroleum Engineering, Clausthal University of Technology, Agricolastr. 10, 38678 Clausthal-Zellerfeld, Germany

S. Buske
TU Bergakademie Freiberg, Akademiestraße 6, 09599 Freiberg, Germany

L. Houpt
Freie Universität Berlin, Fachrichtung Geophysik, Malteserstrasse 74-100, 12249 Berlin, Germany

S. Schmitz
DBI Gastechnologisches Institut gGmbH, Fachgebiet Öl/Gasförderung & Gasspeicherung, Halsbrücker Str. 34, 09599 Freiberg, Germany

M. Kühn and U. Münch (eds.), *CLEAN: CO_2 Large-Scale Enhanced Gas Recovery in the Altmark Natural Gas Field*, Advanced Technologies in Earth Sciences,
DOI 10.1007/978-3-642-31677-7_7, © Springer-Verlag Berlin Heidelberg 2013

recovery (no EGR effect). Within the pilot project period of 2 years there would be no CO_2 arrival at the production well located at about 1,600 m from the injector. Penetration of the CO_2 into the reservoir was maximum approximately 800 m and minimum around 250 m depending on the permeability of the formation layer. The gas mixing zone reaches the observation well which is located at about 660 m from the injector already before 1 year of injection.

Under the simulated conditions direct monitoring of the CO_2 front with active seismic measurements will not be feasible as the reservoir layers are too thin and the effects of CO_2 gas replacing reservoir gas are too small. Seismic monitoring has to concentrate on the detection of possible leakages in shallower aquifers where larger velocity and density changes would occur.

7.1 Theoretical Test Case Due to Lack of CO_2 Injection

CLEAN was a scientific programme in support of the pilot Enhanced Gas Recovery (EGR) project of the site owner and operator GDF SUEZ E&P Deutschland GmbH (GDF SUEZ). Under the pilot project, it had been planned to inject nearly 100,000 t of CO_2 into the Altensalzwedel block (Kühn et al. 2011, 2012; Chap. 1). However, the permitting procedure required under mining law was delayed and the responsible mining authority (State Office for Geology and Mining of Saxony Anhalt) did not issue a permit for CO_2 injection during the period of the project. Therefore a test case was studied of the theoretical injection of CO_2 which is presented here. The outlined results are based on a cooperative effort of large parts of the project partners. Hence, many co-authors contributed.

Within the workflow to be shown below the owner GDF SUEZ re-assessed all available data from the reservoir itself. These were mainly production data as well as data gained from logging activities in the wells. All the data were reprocessed, digitized and re-interpreted. To provide reliable and consistent data sets, data were homogenized and harmonized. E.g. the re-interpretation of seismic data was applied to interpret the structural static reservoir models anew.

7.2 Static 3D Geological Modelling

The Altmark gas field is located at the north-western edge of Saxony-Anhalt, Germany, near to the border of Lower Saxony. It is one of a greater number of gas bearing reservoirs of Rotliegend age, which is following a zone from the Netherlands over Northern Germany towards Poland. Besides Groningen gas field in the Netherlands the Altmark field is one of the biggest in Europe. It was discovered in 1968 and still productive until today with more than 420 wells reaching the top reservoir of which about 250 wells are production wells. Today, it is a mature and almost depleted gas field with a large storage potential for carbon dioxide.

Geologically, the Altmark gas field is part of the Southern Permian Basin and belongs to the southern edge of the North East German Basin (NEGB). During the Upper Rotliegend the sedimentation of red coloured sandstones and siltstones as reservoir rocks intercalated by shale layers and less permeable siltstones as non-reservoir rock took place. The source of the clastic material of this multi-layer sequence was located south of the Altmark and is part of the Variscides (Flechtinger Scholle, Scholle von Calvörde). The Rotliegend sedimentary sequence with a total thickness of several hundred metres contains up to 15 reservoir horizons. The sandstone show fairly good reservoir parameters with porosity values up to 26 % and permeability values up to 200 mD (ca. $0.2 \cdot 10^{-12}$ m^2). The hydrocarbon filling migrated from Upper Carboniferous source rocks into the Upper Permian reservoir rocks, which are sealed by the intercalated claystones, Zechstein anhydrites and at least several hundred metres of massive Zechstein salt. Altensalzwedel as mentioned is part of the whole Altmark gas field made of a siliciclastic multilayer reservoir, i.e. reservoir layers (sand and coarse siltstone) are intercalated by non-reservoir layers (siltstone and shale). For the modelling process described here it was decided to differentiate between the static geological model and the dynamic simulation model (Chap. 4).

The area covered by the geological model is approx. 195 km^2 (15 km · 13 km) which is about 1/5 of the Altmark area. The simulation model is following the regional geology and therefore striking from NW to SE, with an extension of approx. 27 km^2 (9 km · 3 km, compare Chap. 3). The average reservoir depth within the models ranges from −3620 m to −3280 m TVD (true vertical depth). Porosity values are going up to 21 % and permeability values up to 390 mD, whereas the mean is approximately 11 mD. The initial reservoir pressure is approx. 42.5 MPa and the reservoir temperature is about 125 °C.

7.2.1 Geological Modelling Workflow

With the use of the commercial modelling software package Petrel[1], a 3D geologic model was constructed. This was preceded by an intensive data re-evaluation campaign. Pre-modelling work was done for the Altensalzwedel area of Altmark leading to creation of a reliable digital database. Elements of the pre-modelling work included screening well and seismic data, digitizing data, correlation of wells and geological mapping. The workflow presented in Fig. 7.1 could then be implemented based on the newly evaluated database with well data and geophysical interpretations. Several loops of parameter optimization were required leading to a static reservoir model containing among others a 3D grid, reservoir properties, faults and well connection data.

The density of seismic information (2D and 3D seismic) for the Altensalzwedel field is good enough. Well data was mostly available in paper format. This required digitizing the data. For each well a petrophysical composite log was created. The composite logs contain stratigraphical information, location of the reservoirs, main logs, a lithological log derived from core description or mud log, the petrophysical measurements and information about perforated and cored intervals.

[1] http://www.slb.com/services/software/geo/petrel.aspx

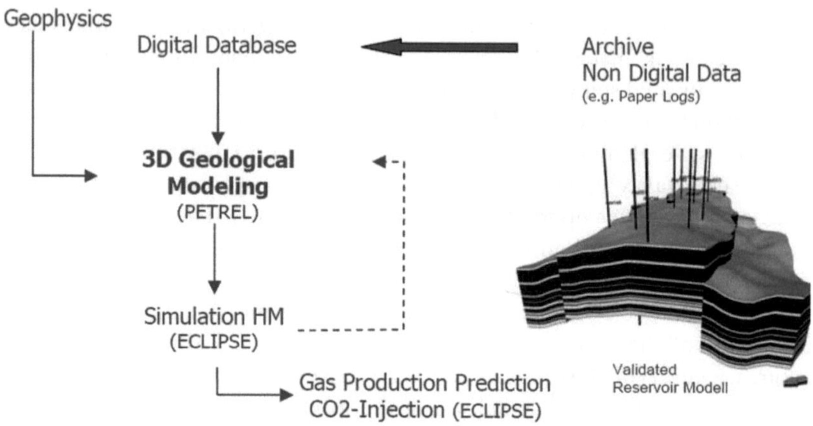

Fig. 7.1 Geological modelling workflow implemented for the Altensalzwedel project

All re-evaluated data (geophysical, geological and reservoir engineering) have been compiled in a project digital database, quality checked and made available for further modelling work. This digital database was also used for well to well correlation, depth control of the log and core data and the construction of different types of well sections to review the results of data preparation.

7.2.2 Optimization of Reservoir Model

As typical for the integrated approach first results of the simulation lead to an update of the geological model (see loop in Fig. 7.1). Permeability multipliers were implemented in the reservoir model and the new property distributions were then given back to the reservoir engineer. Furthermore, simulation faults, faults following the grid geometry and without vertical displacement (transmissibility barriers), were implemented in the reservoir model.

7.3 Dynamic Reservoir Modelling

After the successful set-up of the geological model involving geological, geophysical and petrophysical evaluations and modelling, a static reservoir model was generated using the reservoir modelling tool Petrel. The static model was then taken for further dynamic reservoir modelling and/or simulation. Generally, injection of carbon dioxide (CO_2) into matured gas fields for storage purposes and to enhance gas recovery requires rigorous reservoir engineering and simulation planning. Dynamic numerical approaches offer possibilities to model such complex flow processes while considering the geology and the physical properties of the reservoir gases. Such numerical reservoir simulation methods have been used for studying subsurface CO_2 storage in different contexts and widely reported in open literature. In the following subsections, more details will be described that handle basic components of the reservoir simulation study for CO_2 injection into the Altmark-Altensalzwedel matured gas field which involves wellbore dynamic modelling, history matching the dynamic model and prediction scenarios for CO_2 injection targeting enhanced gas recovery. The numerical computations are performed using the Schlumberger Geo-Quest all-purpose industry commercial Black-oil and full compositional simulators ECLIPSE 100 and ECLIPSE 300[2] respectively.

7.3.1 Wellbore Modelling

Wellbore modelling was performed for the main injector (Well S13, compare Chap. 4) to the depth of approximately 3,500 m. The ultimate goal was to identify parameters that will ensure effective fluid injection in the reservoir and stable as well as effective flow of the CO_2 along the wellbore at the same time. The commercial software package ECLIPSE-VFPi[3] was used in the process. An alternative modelling was performed with BOCO, an in-house product of DBI gGmbH (compare Chap. 3). During the modelling effort, the main parameters and the major considerations that could affect the injection of CO_2 in a depleted gas reservoir were closely examined. It came out that appropriate wellbore modelling would be possible only if the initial and boundary conditions imposed by the wellhead injection parameters (pressure, temperature and rate) and by the reservoir (injectivity index) and wellbore surrounding rock respectively were integral part of the modelling process. This has also been recognized by Carroll and Maddocks (1999) in their work on modelling acid gas injection and by Paterson et al. (2008) and Lu and Connell (2008) in their works on numerical modelling of temperature and pressure profiles of CO_2 in wellbores. Consequently, the components of the injection system (wellhead, borehole and surrounding rock and reservoir) were analysed for their influences on the effectiveness of the flow in the wellbore and the resulting bottom hole temperature (BHT) and bottom hole pressure (BHP).

Bottom hole pressure and temperature vary as a function of the injection rate. The higher the rate, the lower the BHP and the BHT. It is further realized that the wellhead temperature has no significant influence on the bottom hole temperature. The main parameter considered for the reservoir is the injectivity index, which determines the rate of fluid intake by the reservoir. In fact the injection rate has to be kept below the maximum rate that the reservoir can take at the resulting bottom hole pressure. Figure 7.2 shows that at a 4.2 MPa (42 bar) bottom hole pressure, the reservoir can take up to 80,000 m^3/day of CO_2.

The surrounding rock interacts thermally with the flowing fluid as the latter descends down the formation. It results in a cooling or heating of the CO_2

[2] http://www.slb.com/services/software/reseng.aspx

[3] http://www.slb.com/services/software/reseng/eclipse_options/vfpi.aspx

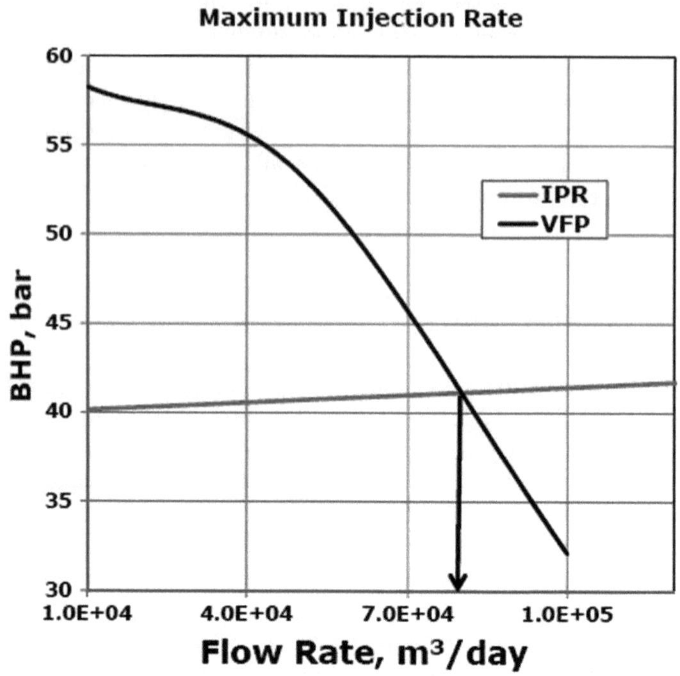

Fig. 7.2 Maximum injection rate for 4.2 MPa (42 bar) bottom hole pressure (*BHP*)

depending on whether the CO_2 or the formation has higher temperature (compared to each other) at the depth of interest. The hot fluid will cool in the first few hundred metres (whenever its temperature is higher than the surrounding rock's temperature) after which, its temperature will start approaching surrounding rock's temperature. This also depends on the flow rate. The lower the rate the closer the fluid temperature will be to the reservoir temperature. The rapid heating of the formation rock by the gas in the first few hundred meters would generate thermal stresses which could loosen the cement bond between formation and casing. This should be avoided. Therefore it is recommended to always maintain a temperature difference of less than 30 °C between the fluid and the surrounding rock (Chap. 3) in the Altmark injection wells.

Another important factor that would constrain the setting of injection parameters is the need to avoid phase change in the wellbore. Its consideration required checking the temperature and pressure evolution along the wellbore with the CO_2 phase diagram to identify if and where phase change will occur.

Figure 7.3 shows the maximum achievable injection rates at the chosen tubing head pressure (THP). These are approximately 80,000 m^3 for 3.5 MPa (35 bar) THP and 110,000 m^3 for 4.0 MPa (40 bar) THP.

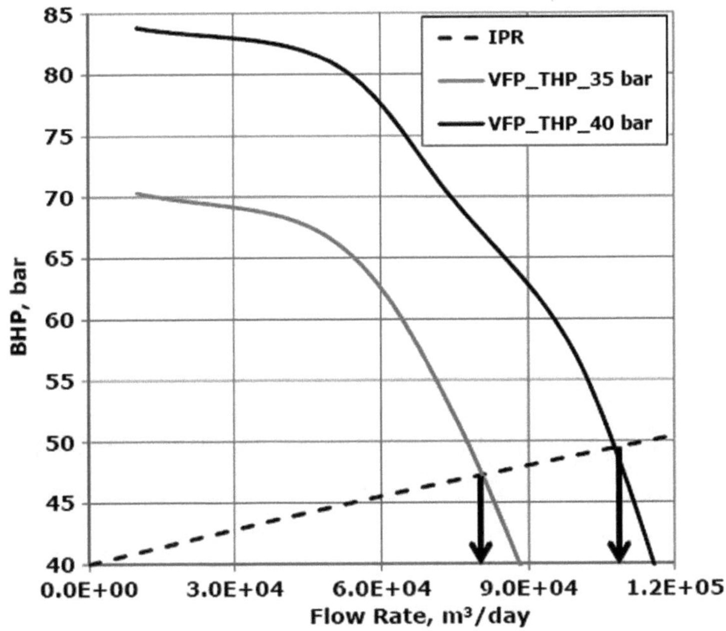

Fig. 7.3 Maximum injection rate at 3.5 MPa (35 bar) and 4.0 MPa (40 bar) THP

The results are in line with the conclusions of Paterson et al. (2008) who reported that for a CO_2 injection well, bottom hole temperature is confined between well-head and reservoir temperature.

It is important to note that keeping injection temperature equal to surface temperature (10 °C) and keeping injection pressure closer to reservoir pressure (35 bars and 40 bars) would help to avoid unnecessary compression to higher pressure and heating to higher temperature, which will cut operation expenditure and help reduce emission of CO_2 to the atmosphere as a result of compression.

7.3.2 Model Validation: History Matching

The production data are available on a daily and monthly basis starting from 1977. Therefore, the simulation was started on a monthly time step until April 1987 and converted to a daily time step onwards. This allowed the development of 31 years of production history. Drill stem tests and pressure build-up, isochronal tests and documented bottom hole and tubing head pressure measurements of the initially perforated wells are inspected and prepared as observed data.

Coupled to the adverse layering of the Altensalzwedel structure variable gas compositions were taken from the gas analysis. Thus the producing reservoir system of the Altensalzwedel wells varies during the production period in a well. Table 7.1 gives the range of the main gas components and quality of the produced

Table 7.1 Gas composition Altmark-Altensalzwedel as used for compositional model

	Mole percent, %			
Component	Min.	Max.	Average	Compositional
CH_4	18	30	25.6	25.88
C_2H_6	0.2	0.4	0.27	
CO_2	0.01	0.9	0.64	0.64
N_2	69	81	73.26	73.48
H_2O, ppm	–	360,000	74,000	74,000

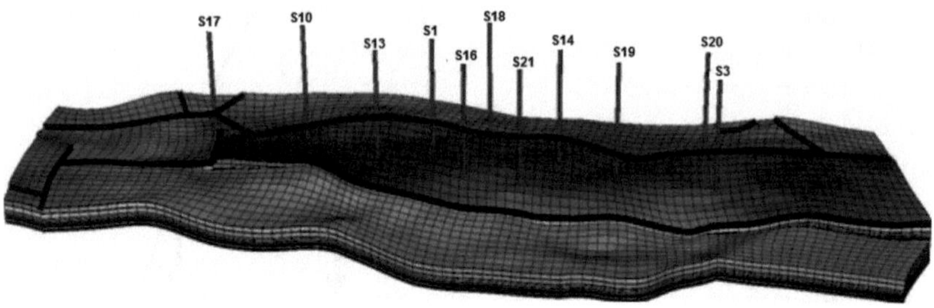

Fig. 7.4 Altensalzwedel simulation model with 11 wells for history matching and major faults from seismic

formation water. For practical reason averaged values of the measured gas compositions were calculated to create representative PVT (pressure, volume, temperature) tables.

The simulation model is discretized into 97 · 26 · 56 elements giving some 141,232 grid blocks of which 117,998 were active. There are a total of 11 wells included with changing perforation settings during the history. Originally, only the main faults worked out by seismic interpretation were built in the 3-D static model from Petrel as shown in Fig. 7.4.

The initial reservoir grip properties of the simulation model are summarized in Table 7.2. In the final history match case a stronger compartmentalization was implemented in the model in order to simulate the production and pressure history of the production wells. These additional barrier systems were partly interpreted in the frame of the geological work. The final results were a number of flow barriers with transmissibility multipliers.

7.3.3 *Injection Prediction*

After the model validation, investigations were performed to predict what will happen if CO_2 is injected into the Altensalzwedel block of the Altmark natural gas field. It is worth mentioning that there is no flowing water present in the model and the residual water saturation were taken from the saturation tables for flow. An injection strategy was predefined to investigate the effect of CO_2 injection on gas recovery and propagation of the CO_2 in the subsurface. The injection strategy

Table 7.2 Initial reservoir grid properties of simulation model

	Value attributes		
Grid property	Max.	Min.	Average
TOPS	3,609.9	3,284.5	3,427.6
DEPTH	3,611.3	3,284.5	3,429.2
DX	144.05	29.116	91.108
DY	213.96	33.447	97.593
DZ	31.503	0.00002	3.3128
NTG	1	1	1
PORO	0.2974	0.00014	0.07316
PORV	12,496	6.142	1,037
kx/kz	0.2999	0.0082	0.3
Sw – Initial	1	0.1462	0.66646
Sg – Initial	0.8538	0	0.33354

followed was to inject 100,000 t of gaseous CO_2 through well S13 over 1.5–2 years and producing methane from well S14 (Fig. 7.4). Well S1 is used as an observation well to monitor the arrival of CO_2. A constant temperature profile was used from 10 °C at the surface increasing with depth to approximately 125 °C at the perforation level of the well. The coupling gives the dependency of the bottom hole pressure on the gas rate for different tubing head pressures. With this injection strategy defined, the base case (production only) and three different scenarios with CO_2 injection were simulated to assess the effect on the reservoir pressure and thus the efficiency of the EGR process and the propagation of the CO_2 through the reservoir:

- Base Case: No injection takes place while gas is produced (continuous depletion). Production was through well S13 at a production rate of 50,000 sm^3/day.
- Scenario 1: Injection of 100,000 t of CO_2 into well S13 over 2 years at a rate of approximately 143 t/day. There was no gas production (pure CO_2 storage).
- Scenario 2: Similar to scenario 1. Difference was gas production from well S14 at a rate of 50,000 sm^3/day. Initially depleted pressure of 4 MPa at S13.
- Scenario 3: Injection of 100,000 t of CO_2 into well S13 over approximately 1.43 years at an injection rate of 200 t/day (maximum injection rate).

With regard to the pressures and temperature conditions, CO_2 was injected in all cases as a gas below its critical temperature and pressure ($p_{crit} = 7.38$ MPa and $T_{crit} = 31.04$ °C). Moreover, all simulations were terminated as soon as the CO_2 concentration in the production stream reached 20 %.

The pressure performance prediction of the average reservoir pressure is given in Fig. 7.5 and the well injection pressures in Fig. 7.6 for the different scenarios. It can be seen that the bottom hole pressure increases on average by 18–23 bar depending on the scenario. Further, the effect (increase of average pressure by about 2 bar) of the tested injection rates on the average reservoir pressure is beyond the simulation accuracy. Under consideration of the volume ratio of injected CO_2 and the gas in

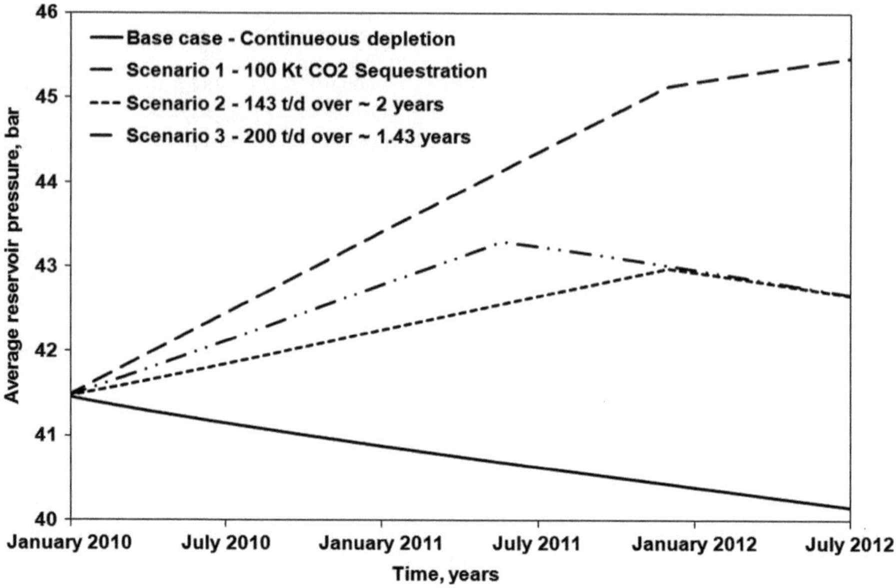

Fig. 7.5 Average reservoir pressure performance prediction in the Altensalzwedel block

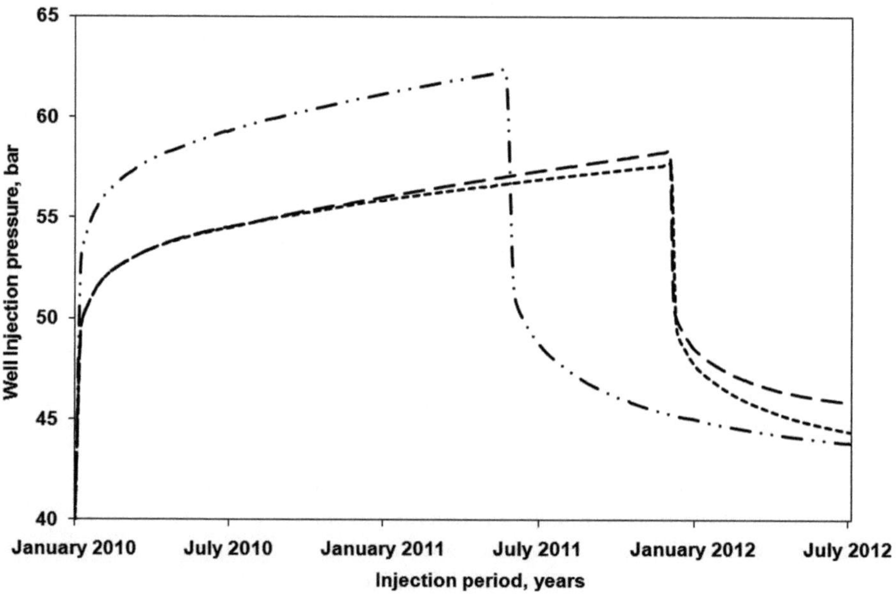

Fig. 7.6 Well injection pressure performance prediction

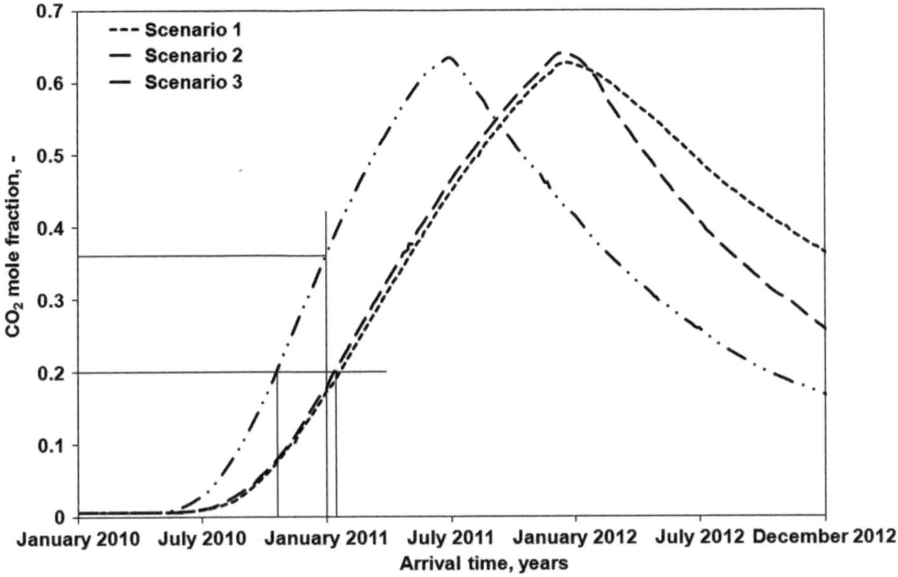

Fig. 7.7 Prediction of time period required for a concentration of 20 % CO_2 to arrive at the observation well S1

place of approximately 0.06, it can be concluded quantitatively that injecting 100,000 t of CO_2 into Altensalzwedel subfield does not increase gas recovery significantly. Hence, there is no EGR effect. This could imply re-planning for higher amounts of CO_2 for the pilot test.

Figure 7.7 shows the increase of the CO_2 concentration with time at the observation well S1 approximately 660 m away from the injection well S13. It can be seen that the injection rate as presented by the different scenarios influences the arrival of different CO_2 concentrations at the observation well.

Figure 7.8 presents the lateral distribution of CO_2 in the reservoir for the eighth layer with the largest flow capacity of $(kh)/\phi = 1{,}172$ after 1 year at the end of injection (2 years of injection). This view shows that the CO_2 distribution is in accordance with geological anomalies. The CO_2 plume clusters around the injection well S13 with a finger along one of the south-east striking faults towards the production well S14. The native natural gas mixes with the injected CO_2 and forms a gas mixing zone seen in white on the aerial view. This mixing zone has a different shape depending on the direction of flow with the mixing in the south-east and north-east directions being highest. The observation well S1 approximately 660 m away from the injector is already in the mixing zone after 1 year of injection. However, the CO_2 front is still far away from the production well S14 and therefore does not influence gas production. Hence, there is no arrival of the CO_2 within the project period of 2 years at the production well and considering the injected amount of CO_2, it would even not occur after 8 years of natural gas production.

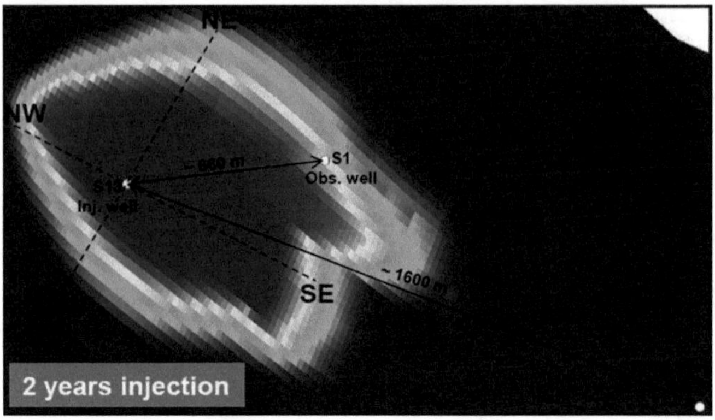

Fig. 7.8 Top view of layer 8 (highest flow capacity, (kh)/ϕ = 1,172) with CO_2 concentration distribution and mixing zone evolution. Note the effect of the fault in the model and the CO_2 tongue towards the production well

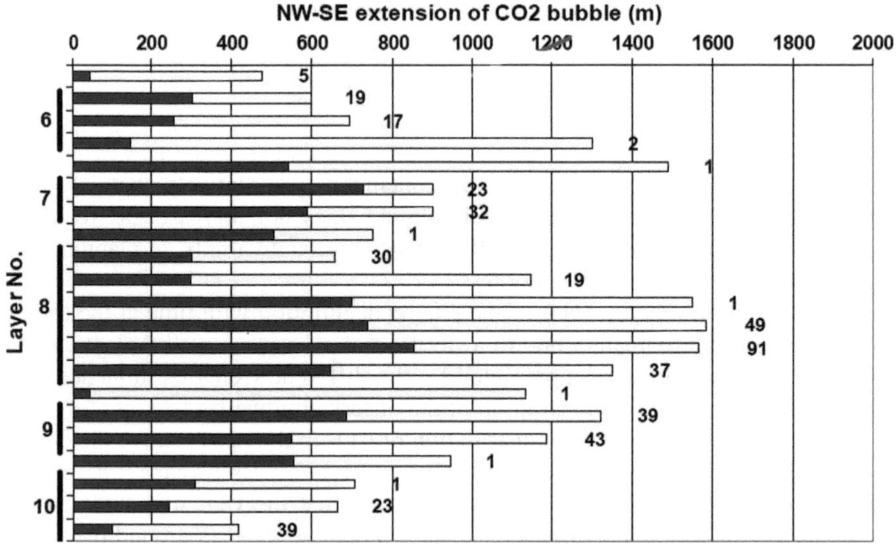

Fig. 7.9 Extension of the CO_2 plume in NW-SE direction for each layer after injection of 100,000 t of CO_2

Figure 7.9 presents a vertical cross section (NW-SE) through the reservoir from the compositional models. This shows how far the CO_2 migrated into the reservoir from the injection for within the various layers of the model. The higher the permeability of a layer the further CO_2 moved into the reservoir.

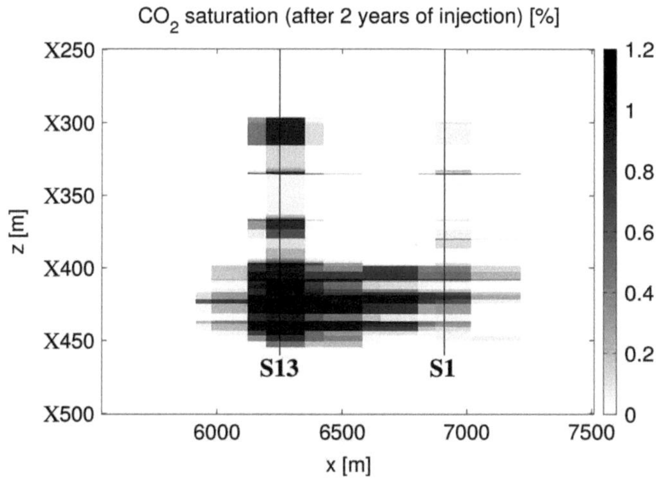

Fig. 7.10 Spatial spreading of simulated CO_2 concentration

7.4 Active Seismic Monitoring

Based on the reservoir simulation results above, location and magnitudes of changes in seismic velocities and density due to 1 and 2 years of CO_2 injection were determined. Based on this information the elastic seismic wave field was calculated with a 2D finite difference code (2D-FD) from Bohlen (2002). Comparing these results with the calculated wave field of the baseline (unperturbed models) the expected changes in the seismic wave field were estimated. To image the changes in the velocity and density models amplitude changes in the seismic wave fields were analysed, which are sensitive to changes in the reflection coefficient, and coda-wave interferometry (Snieder et al. 2002) was used, which is sensitive to time shifts.

7.4.1 Seismic Wave Field Simulation

To calculate the velocity changes and to build the model for the 2D-FD wave field simulations the following steps were carried out:

- In a first step the simulated saturation and pressure values on the plane spanned by the injection well (S13) and monitoring well (S1) were taken. This is the direction from the observation well, where the largest velocity and density changes were expected. Due to the fact that the simulation grid was oriented along the reservoir structures but for the FD simulations a regular grid was needed, the reservoir layers were flattened in the vertical direction. This flattening is feasible, as the vertical variations of the reservoir layers do not exceed 16 m on a distance of 2,000 m. Finally this provided distributions of pore pressure and saturation as it is shown in Fig. 7.10 (after 2 years of CO_2 injection).

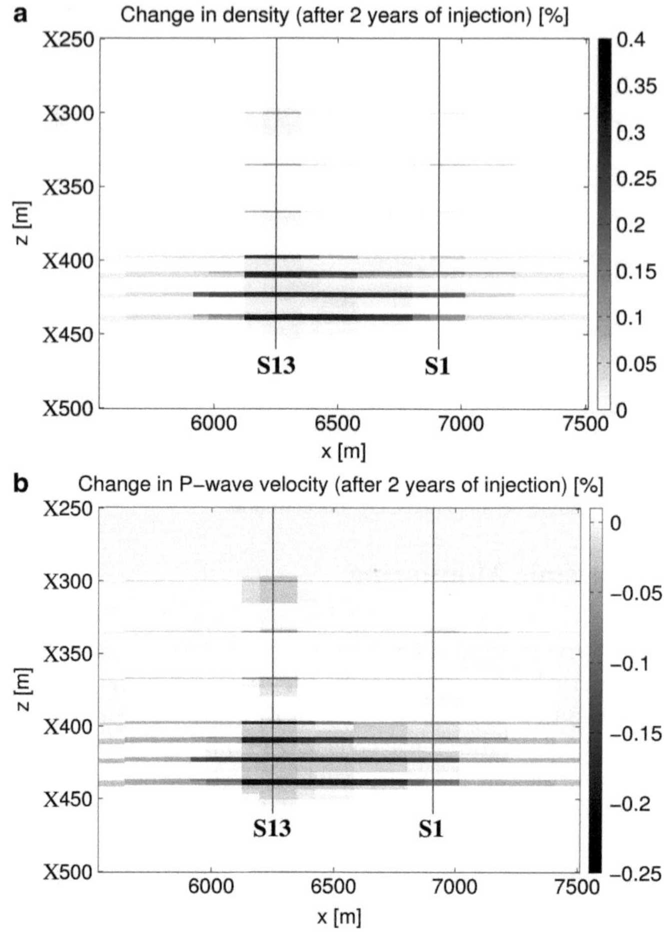

Fig. 7.11 Distribution of (**a**) density and (**b**) velocity changes of the reservoir rock saturated with CO_2 after 2 years of injection compared to the initial reservoir conditions

- In the second step the respective velocity and density change of each grid cell was calculated by use of the saturation and pressure conditions before, after 1 year and after 2 years of injection. Therefore the same steps as described in Chap. 5 were conducted. Figure 7.11 shows the calculated velocity changes after 2 years of injection compared to the initial conditions. Largest velocity changes are much less extended than the high saturation zones in Fig. 7.10. This is due to the low porosity of the regions between the reservoir layers. In these interlayers the porosity is set to 1 % while the porosity of the reservoir layers is up to 20 %. This causes very different effects in saturated rock properties for the same CO_2 saturation values.

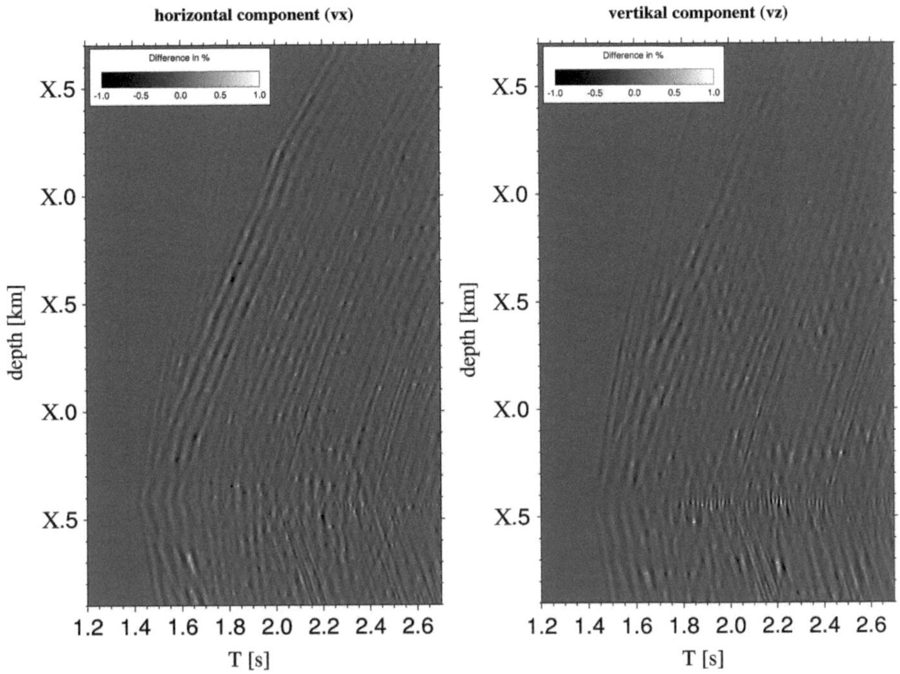

Fig. 7.12 Differences of seismograms calculated from models of the baseline and after 2 years of injection for an offset of 2,900 m

The change in S-wave velocity results from the change in density only, as the shear modulus μ_{sat} is not affected by pore fluids. Finally these changes were implemented in an existing velocity and density model which was based on existing logging data from the monitoring well. For the unperturbed model (without changes) and the perturbed models (changes in seismic properties after 1 and 2 years of injection) the seismic wave field was calculated for different source positions (offset of 2,900 and 5,850 m) and a hypothetical 390-level 3C-receiver array in the observation well with an inter receiver distance of 10 m. Here the 2D-FD code from Bohlen (2002) was used.

7.4.2 Amplitude Changes

The change in reflection coefficients, which are affected by velocity and density changes, can be seen in a change in the seismograms and here especially in a change in the amplitudes of the reflected waves. They were analysed by calculating the difference of the seismic sections calculated for the velocity and density models of the initial reservoir conditions and the conditions after 1 and 2 years of injection. Figure 7.12 shows the difference in the horizontal (left) and vertical (right) seismogram sections calculated after 2 years of injection. The source and receiver offset is 2,900 m. The amplitude changes in both components are very low (maximum at

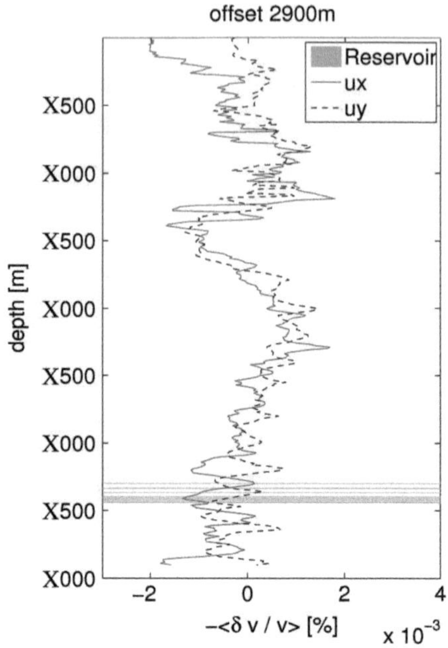

Fig. 7.13 Mean velocity changes calculated from time shifts in the seismic wave filed of the baseline and the repeat after 2 years of injection

about 1 %) and far below the limiting value of 5 %, which is assumed for the borehole seismometers.

7.4.3 Coda Wave Interferometry

A change in velocity will produce shifts in the arrival times of wavelets. These changes were calculated with coda wave interferometry and with this mean velocity changes were estimated. The procedure is explained in more detail in Zhou et al. (2010). Figure 7.13 shows the results for the seismic sections after 2 years of injection for the horizontal (u_x) and vertical component (u_y). The mean velocity changes are below the assumed threshold of 0.01 %. Especially in the depth range of the reservoir layers no increase in the mean velocity curves can be found. This is due to the low velocity changes in the reservoir and the low spatial expansion.

References

Bohlen T (2002) Parallel 3-D viscoelastic finite difference seismic modeling. Comput Geosci 28:887–899

Carroll JJ, Maddocks JR (1999) Design consideration for acid gas injection (online). Presented at the Lawrance Reid gas conditioning conference, Norman. Available at http://www.gasliquids.com/papers/lrgcc99.pdf

Kühn M, Förster A, Großmann J, Meyer R, Reinicke K, Schäfer D, Wendel H (2011) CLEAN: preparing for a CO_2-based enhanced gas recovery in a depleted gas field in Germany. Energy Proc 4:5520–5526. doi:10.1016/j.egypro.2011.02.538

Kühn M, Tesmer M, Pilz P, Meyer R, Reinicke K, Förster A, Kolditz O, Schäfer D, CLEAN Partners (2012) CLEAN: CO2 large-scale enhanced gas recovery in the Altmark natural gas field (Germany): project overview. Environ Earth Sci. 67(2):311–321 doi:10.1007/s12665-012-1714-z

Lu M, Connell LD (2008) Non-isothermal flow of carbon dioxide in injection wells during geological storage. Int J Greenhouse Gas Control 2:248–258

Paterson L, Lu M, Connell L, Ennis-King JP (2008) Numerical modeling of pressure and temperature profiles including phase transition in carbon dioxide wells. SPE Annu Tech Conf Proc 4:2693–2703

Snieder R, Grêt A, Douma H, Scales J (2002) Coda wave interferometry for estimating nonlinear behavior in seismic velocity. Science 295:2253–2255

Zhou R, Huang L, Rutledge JT, Fehler M, Daley TM, Majer EL (2010) Coda-wave interferometry analysis of time-lapse VSP data for monitoring geological carbon sequestration. Int J Greenhouse Gas Control 4:679–686

Chapter 8
Public Outreach

Andreas Hübner, Tanja Kollersberger, Peter Pilz, Maja Tesmer and Michael Kühn

Abstract Critical public perception of capture and subsequent geological storage of CO_2 (CCS – Carbon Capture and Storage) may turn out to be a hindrance to the development of the CCS technology. Therefore, in addition to the scientific and technological development, acceptance by the public is a decisive prerequisite for the successful implementation of this technology. Notwithstanding the result of any decision, it is desirable that decisions on acceptance or rejection of a technology are based on factual arguments rather than on speculations. Public outreach within the CLEAN project therefore aimed to support opinion-forming based on scientific knowledge and transparent discussion of concerns, benefits and knowledge gaps. All this was addressed with a variety of media, tailored to the respective target audience. However, conducting effective public outreach does not necessarily guarantee the success of a project.

8.1 Introduction

The Federal Republic of Germany considers capture and subsequent geological storage of CO_2 (CCS – Carbon Capture and Storage) to be a promising key technology to reduce CO_2 emissions into the atmosphere. However, critical public perception of CCS may turn out to be a hindrance to the development of CCS (TAB 2008). Therefore, in addition to the scientific and technological development of CCS, public acceptance is a decisive prerequisite for the successful implementation of this technology.

Public outreach within the CLEAN project aimed to support opinion-forming of the various stakeholders based on factual arguments rather than on speculation. Concerns regarding the geological storage of CO_2 in general and the CLEAN

A. Hübner • T. Kollersberger • P. Pilz • M. Tesmer • M. Kühn (✉)
Helmholtz Centre Potsdam, GFZ – German Research Centre for Geosciences, Telegrafenberg, 14473 Potsdam, Germany
e-mail: michael.kuehn@gfz-potsdam.de

M. Kühn and U. Münch (eds.), *CLEAN: CO_2 Large-Scale Enhanced Gas Recovery in the Altmark Natural Gas Field*, Advanced Technologies in Earth Sciences,
DOI 10.1007/978-3-642-31677-7_8,

project in particular were addressed and put into a scientific perspective. Public outreach was based on a transparent, balanced, and citizen-friendly information policy. Knowledge gaps and the potential risks of geological CO_2 storage were addressed as well as the potential benefits of enhanced gas recovery with CO_2.

At the beginning of the CLEAN project, the general public had a large information deficit with regard to CCS. A survey revealed that the processes related to CCS were largely unknown to the public (Donnermeyer 2009). Results also suggested that survey participants perceived capturing CO_2 to be less problematic than transport and geological storage of CO_2.

8.2 Public Relation Concept and Exhibits

The GFZ – German Research Centre for Geosciences has acquired considerable experience in public outreach related to geological CO_2 storage at the Ketzin pilot site (Schilling et al. 2009), which has been operated by the GFZ since 2004 (Martens et al. 2012). Based on these experiences, a communication strategy for CLEAN was developed at the beginning of the project and was agreed on by the project's steering committee. The communication strategy identified the following core elements: (1) establishing a local information centre, (2) set up a project website and (3) conduct local informational briefings open to the public. An important requirement was that outreach material should be tailored to the audiences and take advantage of different media. Independent research on the public response and perception of these measures or the development of public acceptance was not part of the CLEAN project.

One of the strategic measures for public outreach identified in the communication concept was to establish a local information centre for the CLEAN project in the town of Salzwedel. However, as a result of the dynamic development of the socio-political debate on geological CO_2 storage at communal and federal levels the information centre was not opened in Salzwedel. Instead, it was decided to integrate the concept and the ideas into the visitor centre which already existed at the Ketzin pilot site in Brandenburg.

8.2.1 Interactive Exhibits

With the interactive exhibits the users are informed about the basic aspects of the CLEAN project. Compared to passive exhibits the advantage of interactive ones is that they motivate additional questions and insights about the complex interrelationships involved in geological CO_2 storage. The exhibits are described below in detail.

8.2.1.1 EGR – Enhanced Gas Recovery Exhibit

The "EGR" exhibit (Fig. 8.1) demonstrates and explains the following basic aspects of the CLEAN project:

Fig. 8.1 General view of the "EGR" exhibit. The height of the exhibit is approx. 120 cm

- The schematic geological cross section of the Altmark region, including the reservoir and the impermeable cap rock (salt and clay rock formations),
- The depth of the natural gas reservoir / CO_2 storage formation,
- The EGR process.

The exhibit shows a generalized geological cross section of the Altmark region down to a depth of 3,500 m, including an injection well, a monitoring well and a production well (Fig. 8.1). On top of the exhibit (above the geological profile), two CO_2 tanks similar to those at the Maxdorf injection facility and the well heads for the displayed injection, monitoring, and production wells are shown. In front of the geological cross section, a button is provided to start the animation of the EGR process. CO_2 injection and associated natural gas production are simulated by means of LEDs. The concept of EGR is explained with a text and a chart shows the subsurface multi-barrier system.

8.2.1.2 Permeability Exhibit

The "Permeability" exhibit (Fig. 8.2) allows users to test the permeability of different rock types with respect to air, i.e., the ability of the rock to allow air to flow through. The users can choose between sandstone, siltstone, salt rock and claystone. The exhibit illustrates key features of a reservoir:

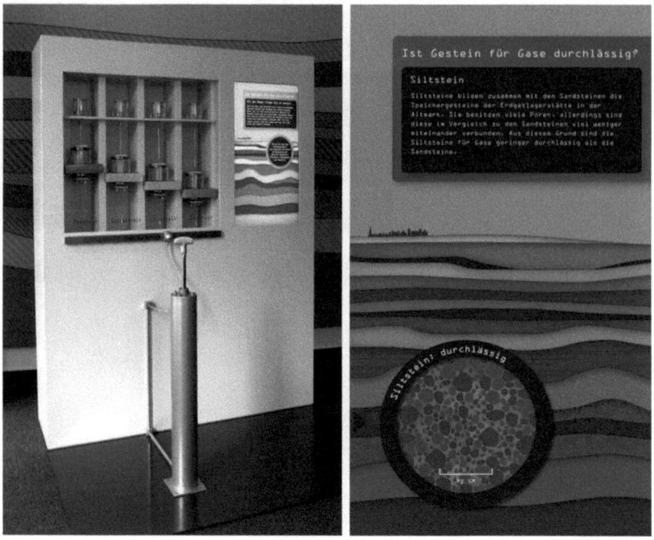

Fig. 8.2 *Left*: general view of the "Permeability" exhibit. The height of the exhibit is approx. 190 cm. *Right*: detail of animated screen

- Permeabilities of different rock types,
- Allocation of different rock types to reservoir or cap rock,
- Spatial arrangement of reservoir and cap rocks and the illustration of a geological barrier.

The user applies pressure to the selected rock type via a manual air pump. For sand- and siltstone, the air flows through the rock and is then led into a container filled with liquid, in which air bubbles demonstrate the movement of the air.

Choosing sandstone, pumping is easy because of its high permeability. With siltstone, on the other hand, the user must clearly use more force due to the lower permeability. Accordingly, fewer bubbles are produced in the liquid container. For salt rock and claystone the user experiences very high resistance at the pump and no bubbles are formed in the respective liquid containers. For these rock types, it gets obvious that there is no through-flow.

The screen to the right of the rock samples provides additional information about the structure and texture of each rock type. In the case of sand- and siltstone, the air that migrates through the pore space of the rock is illustrated by small triangles. Depending on the pumping strength, these triangles move with different velocities from bottom to top through the pore space and by doing so they mimic the gas flow through the rock. For claystone and salt rock, no gas flow due to the extremely low permeability of these rocks is shown on the screen.

8.2.1.3 Information Terminal Exhibit

An information terminal (Fig. 8.3) was designed and built to provide project-specific and general information and enable visitors to inform themselves easily.

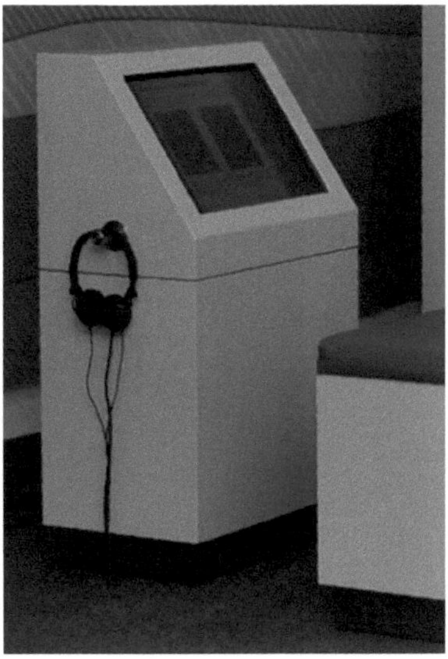

Fig. 8.3 General view of the "Information terminal" exhibit. The height of the exhibit is approx. 85 cm

Its advantage is that the interested user may find a large range of information and answers to questions with little effort.

The exhibit consists of a computer with a touch screen and a headphone on either side of the cabinet, as well as a bench on which two users may easily sit side by side. By way of a PowerPoint Presentation in kiosk mode, a selection of text, charts, films and audio files provide the user with information about the CLEAN project, and general information about research on the geological storage of CO_2.

8.2.2 Further Information and Demonstration Material

8.2.2.1 Permeability Experiment

The physics of two-phase fluid flow govern the hydraulic processes during the injection of CO_2 into saline aquifers. An experimental setup was designed to explain the fundamentals of the relevant flow laws, using air as substitute for CO_2 (Fig. 8.4). The central part of the experiment is formed by a cylindrical sandstone core in a transparent pressure casing. Experiments can be carried out with three different saturation states. The mobile medium is gas, while the core sample may be either fully water saturated (wet), dry or partly saturated, i.e., containing residual water. The experiments allow the determination of intrinsic and relative permeability, entry pressure and pore radius.

8.2.2.2 Rock Cubes Experiment

A major retention mechanism during geological CO_2 storage is the structural trapping of the gas by very low-permeability rock layers which act as a barrier.

Fig. 8.4 The sample cylinder, which is the central part of the experiment with sandstone sample. Length of rock sample: 10 cm

This process is demonstrated with a simple, easy-to-transport exhibit involving two different 5-cm cubes of rock: one of sandstone and one of claystone. The cubes are placed side by side in a Petri dish filled with water. The water is drawn into the pore spaces of the sandstone and retained there by capillary forces. For the claystone, however, no water is drawn in and thus no permeability is observed.

8.2.2.3 Salt Exhibit

The Zechstein salt rock formation which is present directly above the Altmark natural gas reservoir rocks plays a prominent role as cap rock. The presence of the salt rocks is an important prerequisite for the general feasibility and safety of CO_2 storage in the Altmark. To demonstrate that rock salt indeed is solid rock, an exhibit was manufactured (Fig. 8.5), with a piece of rock salt approx. 40 cm in width mounted on a base. A short text explains the formation of the Zechstein salt rocks.

8.2.3 Films

The CLEAN project was presented in a broader context by way of a film series entitled "The geological storage of CO_2" (Fig. 8.6), made up of six approximately 5-min short film sequences. The titles of the six film sequences are:

- "Forschung für den Klimaschutz" (Research for Climate Protection),
- "Geologische Voraussetzungen und Kapazitäten" (Geological Conditions and Capacities),
- "Rückhaltemechanismen" (Trapping Mechanisms),

Fig. 8.5 View of the rock sample of the "Salt" exhibit. The rock has a width of approx. 40 cm

Fig. 8.6 Filming laboratory experiments at TU Clausthal-Zellerfeld

- "Speicherüberwachung und Prognose" (Monitoring and Prediction),
- "Pilotstandort Ketzin" (Pilot Site Ketzin),
- "Forschungsprojekt CLEAN in der Altmark" (Research Project CLEAN in the Altmark).

The films were produced to explain the complexity of the topic in a simple and appealing way without lacking scientific accuracy. The sequences were kept to a reasonable length appropriate for the use on a website and with the information terminal exhibit. One of the sequences deals explicitly with CO_2/EGR technology

Fig. 8.7 Screenshot of the home page of the CLEAN project website (http://www.clean-altmark.org)

related to the Altmark project site. All sequences can be viewed on the project website of the Ketzin pilot site[1] and are accessible via the homepage of the GFZ – German Research Centre for Geosciences.[2]

8.2.4 *Website*

A web-based information forum was created[3] to provide detailed information about the basics of the EGR technology and CO_2 storage in general as well as the latest research developments of the CLEAN project (Fig. 8.7).

In addition, this website fulfilled a second function by providing a work space for the partners of the CLEAN project. Authorized project members may access the data management system (DMS) of the CLEAN project or the "Information platform", a web-based project management tool of the CLEAN project.

---

[1] http://www.co2ketzin.de/

[2] http://www.gfz-potsdam.de/portal/gfz/Struktur/GeoEngineering-Zentren/CO2-Speicherung

[3] http://www.clean-altmark.org

8.3 Presentations and Press Relation

Communication with the regional population as well as with the oppositional local action group took place through various public information events and one exhibition. Furthermore, a series of informational meetings were held with decision makers such as the mayor and city council of Salzwedel, as well as the district administrator and environmental committee of the Altmark district.

In addition, press releases were compiled, interviews were given, and replies to inquiries from the print media, radio and TV stations, and free-lance journalists were provided. News coverage on the CLEAN project was documented and archived. During the lifetime of the project 9 TV and 13 radio broadcasts as well as 536 newspaper articles related to the CLEAN project.

8.4 Conclusions

Public interest in geological CO_2 storage projects is high, particularly in the potential target areas. Therefore, in research projects where large-scale field work is involved, sufficient personnel and financial resources are required for public outreach. It is not possible for the scientists participating in a project to fulfil this major task additionally to their research work. Public outreach should be an integral part of the project management and it requires careful planning and implementation. The distribution of factual and transparent information about the research motivation and objectives is an important measure to support public opinion-forming. However, conducting effective public outreach does not necessarily guarantee the success of a project.

Predictions of the opinions and interests of the stakeholders are possible only to a limited extend. For example, election campaigns or vociferous public opposition may influence political decision makers and diverging views of industry and science might change the way of implementation of public outreach in a project. Consistent communication within the project was defined in the project agreement to ensure information and knowledge transfer from the project to the public and any other stakeholder. This was necessary, as the project was publicly discussed in the overall context of climate and energy policy and the possible introduction of the CO_2/EGR technology in Germany.

References

Donnermeyer M (2009) CCS – Eine Technologie im Klimaschutzportfolio. Workshop Institut für Energie- und Regulierungsrecht, 27 Mai 2009, presentation. http://www.enreg.de/content/material/2009/27.05.2009.Donnermeyer.IZKLIMA.pdf. Available 4 May 2012

Martens S, Kempka T, Liebscher A, Lüth S, Möller F, Myrttinen A, Norden B, Schmidt-Hattenberger C, Zimmer M, Kühn M, Ketzin Group (2012) Europe's longest-operating on-shore CO_2 storage site at Ketzin, Germany: a progress report after three years of injection. Environ Earth Sci. 67(2):323–334 doi:10.1007/s12665-012-1672-5

Schilling FR, Ossing F, Würdemann H, CO_2SINK Team (2009) Public acceptance for geological CO_2-storage. Geophys Res Abstr 11, EGU2009-13801-2

TAB – Büro für Technikfolgen-Abschätzung beim Deutschen Bundestag (2008) Technikfolgen-abschätzung (TA) CO_2-Abscheidung und -Lagerung bei Kraftwerken. Sachstandsbericht zum Monitoring "Nachhaltige Energieversorgung". Bericht des Ausschusses für Bildung, Forschung und Technikfolgenabschätzung, Drucksache 16/9896. http://dip21.bundestag.de/dip21/btd/16/098/1609896.pdf. Available 4 May 2012

Index

M. Kühn and U. Münch (eds.), *CLEAN: CO_2 Large-Scale Enhanced Gas Recovery in the Altmark Natural Gas Field*, Advanced Technologies in Earth Sciences,
DOI 10.1007/978-3-642-31677-7,

MIX
Papier aus verantwortungsvollen Quellen
Paper from responsible sources
FSC® C105338

If you have any concerns about our products,
you can contact us on
ProductSafety@springernature.com

In case Publisher is established outside the EU,
the EU authorized representative is:
Springer Nature Customer Service Center GmbH
Europaplatz 3, 69115 Heidelberg, Germany

Printed by Libri Plureos GmbH
in Hamburg, Germany